T0182062

Transients for Electrical Engineers

Oliver Heaviside (1850–1925), the patron saint (among electrical engineers) of transient analysts. (Reproduced from one of several negatives, dated 1893, found in an old cardboard box with a note in Heaviside's hand: "The one with hands in pockets is perhaps the best, though his mother would have preferred a smile.")

Frontispiece photo courtesy of the Institution of Electrical Engineers (London)

Paul J. Nahin

Transients for Electrical Engineers

Elementary Switched-Circuit Analysis in the Time and Laplace Transform Domains (with a touch of MATLAB®)

Foreword by John I. Molinder

Paul J. Nahin
University of New Hampshire
Durham, New Hampshire, USA

ISBN 978-3-030-08490-5 ISBN 978-3-319-77598-2 (eBook)
https://doi.org/10.1007/978-3-319-77598-2

Printed on acid-free paper

This Springer imprint is published by the registered company Springer International Publishing AG part of Springer Nature.
The registered company address is: Gewerbestrasse 11, 6330 Cham, Switzerland

"An electrical transient is an outward manifestation of a sudden change in circuit conditions, as when a switch opens or closes . . . The transient period is usually very short . . . yet these transient periods are extremely important, for it is at such times that the circuit components are subjected to the greatest stresses from excessive currents or voltages . . . it is unfortunate that many electrical engineers have only the haziest conception of what is happening in the circuit at such times. Indeed, some appear to view the subject as bordering on the occult."

and

"The study of electrical transients is an investigation of one of the less obvious aspects of Nature. They possess that fleeting quality which appeals to the aesthetic sense. Their study treats of the borderland between the broad fields of uniformly flowing events . . . only to be sensed by those who are especially attentive."

—words from two older books[1] on electrical transients, expressing views that are valid today.

---

[1]The first quotation is from Allen Greenwood, *Electrical Transients in Power Systems*, Wiley-Interscience 1971, and the second is from L. A. Ware and G. R. Town, *Electrical Transients*, Macmillan 1954.

In support of many of the theoretical calculations performed in this book, software packages developed by The MathWorks, Inc. of Natick, MA, were used (specifically, MATLAB® 8.1 Release 2013a and Symbolic Math Toolbox 5.10), running on a Windows 7 PC. This software is now several releases old, but all the commands used in this book work with the newer versions and are likely to continue to work for newer versions for several years more. The MathWorks, Inc., does not warrant the accuracy of the text of this book. This book's use or discussions of MATLAB do not constitute an endorsement or sponsorship by The MathWorks, Inc., of a particular pedagogical approach or particular use of MATLAB, or of the Symbolic Math Toolbox software.

To the Memory of
Sidney Darlington (1906–1997)
a pioneer in electrical/electronic circuit analysis,[2] *who was a colleague and friend for twenty years at the University of New Hampshire. Sidney's long and creative life spanned the eras of slide rules to electronic computers, and he was pretty darn good at using both.*

[2]Sidney received the 1945 Presidential Medal of Freedom for his contributions to military technology during World War II, and the 1981 I.E. Medal of Honor. One of his *minor* inventions is the famous, now ubiquitous *Darlington pair* (the connection of two "ordinary" transistors to make a "super" transistor). When I once asked him how he came to discover his circuit, he just laughed and said, "Well, it wasn't *that* hard—each transistor has just three leads, and so there really aren't a lot of different ways to hook two transistors together!" I'm still not sure if he was simply joking.

Foreword

Day and night, year after year, all over the world electrical devices are being switched "on" and "off" (either manually or automatically) or plugged in and unplugged. Examples include houselights, streetlights, kitchen appliances, refrigerators, fans, air conditioners, various types of motors, and (hopefully not very often) part of the electrical grid. Usually, we are only interested in whether these devices are either "on" or "off" and are not concerned with the fact that switching from one state to the other often results in the occurrence of an effect (called a transient) between the time the switch is thrown and the desired condition of "on" or "off" (the steady state) is reached. Unless the devices are designed to suppress or withstand them, these transients can cause damage to the device or even destroy it.

Take the case of an incandescent light bulb. The filament is cold (its resistance is low) before the switch is turned on and becomes hot (its resistance is much higher) a short time after the switch is turned on. Assuming the voltage is constant the filament current has an initial surge, called the inrush current, which can be more than ten times the steady state current after the filament becomes hot. Significant inrush current can also occur in LED bulbs depending on the design of the circuitry that converts the alternating voltage and current from the building wiring to the much lower direct voltage and current required by the LED. Due to the compressor motor, a refrigerator or air conditioner has an inrush current during startup that can be several times the steady state current when the motor is up to speed and the rotating armature produces the back EMF. It's very important to take this into account when purchasing an emergency generator. The inrush current of the starter motor in an automobile explains why it may run properly with a weak battery but requires booster cables from another battery to get it started.

In this book, Paul Nahin focuses on electrical transients starting with circuits consisting of resistors, capacitors, inductors, and transformers and culminating with transmission lines. He shows how to model and analyze them using differential equations and how to solve these equations in the time domain or the Laplace transform domain. Along the way, he identifies and resolves some interesting apparent paradoxes.

Readers are assumed to have some familiarity with solving differential equations in the time domain but those who don't can learn a good deal from the examples that are worked out in detail. On the other hand, the book contains a careful development of the Laplace transform, its properties, derivations of a number of transform pairs, and its use in solving both ordinary and partial differential equations. The "touch of Matlab" shows how modern computer software in conjunction with the Laplace transform makes it easy to solve and visualize the solution of even complicated equations. Of course, a clear understanding of the fundamental principles is required to use it correctly.

As in his other books, in addition to the technical material Nahin includes the fascinating history of its development, including the key people involved. In the case of the Trans-Atlantic telegraph cable, you will also learn how they were able to solve the transmission line equations without the benefit of the Laplace transform.

While the focus of the book is on electrical transients and electrical engineers, the tools and techniques are useful in many other disciplines including signal processing, mechanical and thermal systems, feedback control systems, and communication systems. Students will find that this book provides a solid foundation for their further studies in these and other areas. Professionals will also learn some things. I certainly did!

Harvey Mudd College, Claremont, CA, USA
John I. Molinder
January 2018

Preface

"There are three kinds of people. Those who like math and those who don't."
—if you laughed when you read this, good (if you didn't, well . . . .)

We've all seen it before, numerous times, and the most recent viewing is always just as impressive as was the first. When you pull the power plug of a toaster or a vacuum cleaner out of a wall outlet, a brief but most spectacular display of what appears to be *fire* comes out along with the plug. (This is very hard to miss in a dark room!) That's an electrical transient and, while I had been aware of the effect since about the age of five, it wasn't until I was in college that I really understood the mathematical physics of it.

During my undergraduate college years at Stanford (1958–1962), as an electrical engineering major, I took all the courses you might expect of such a major: electronics, solid-state physics, advanced applied math (calculus, ordinary and partial differential equations, and complex variables through contour integration[3]), Boolean algebra and sequential digital circuit design, electrical circuits and transmission lines, electromagnetic field theory, and more. Even assembly language computer programming (on an IBM650, with its amazing rotating magnetic drum memory storing a grand total of ten thousand bytes or so[4]) was in the mix. They were all great courses, but the very best one of all was a little two-unit course I took in my junior year, meeting just twice a week, on electrical transients (EE116). That's when I learned what that 'fire out of a wall outlet' was all about.

Toasters are, basically, just *coils* of high-resistance wire specifically designed to get red-hot, and the motors of vacuum cleaners inherently contain *coils* of wire that generate the magnetic fields that spin the suction blades that swoop up the dirt out of

[3]Today we also like to see matrix algebra and probability theory in that undergraduate math work for an EE major, but back when I was a student such "advanced" stuff had to wait until graduate school.

[4]A modern student, used to walking around with dozens of *giga*bytes on a flash-drive in a shirt pocket, can hardly believe that memories used to be that small. How, they wonder, did anybody do anything useful with such pitifully little memory?

your rug. Those coils are *inductors*, and inductors have the property that the current through them can't change instantly (I'll show you *why* this is so later; just accept it for now). So, just before you pull the toaster plug out of the wall outlet there is a pretty hefty current into the toaster, and so that same current "wants" to still be going into the toaster *after* you pull the plug. And, by gosh, just because the plug is out of the outlet isn't going to stop that current—it just keeps going and *arcs* across the air gap between the prongs of the plug and the outlet (that arc is the "fire" you see). In dry air, it takes a voltage drop of about 75,000 volts to jump an inch, and so you can see we are talking about impressive voltage levels.

The formation of transient arcs in electrical circuits is, generally, something to be avoided. That's because arcs are *very* hot (temperatures in the thousands of degrees are not uncommon), hot enough to quickly (in milliseconds or even microseconds) melt switch and relay contacts. Such melting creates puddles of molten metal that sputter, splatter, and burn holes through the contacts and, over a period of time, result in utterly destroying the contacts. In addition, if electrical equipment with switched contacts operates in certain volatile environments, the presence of a hot transient switching arc could result in an explosion. In homes that use natural or propane gas, for instance, you should *never* actuate an electrical switch of any kind (a light switch, or one operating a garage door electrical motor) if you smell gas, or even if only a gas leak detector alarm sounds. A transient arc (which might be just a tiny spark) may well cause the house to blow-up!

However, not *all* arcs are "bad." They are the basis for arc welders, as well as for the antiaircraft searchlights you often see in World War II movies (and now and then even today at Hollywood events). They were used, too, in early radio transmitters, before the development of powerful vacuum tubes,[5] and for intense theater stage lighting (the "arc lights of Broadway"). Automotive ignition systems (think *spark plugs*) are essentially systems in a *continuous* state of *transient* behavior. And the high-voltage impulse generator invented in 1924 by the German electrical engineer Erwin Marx (1893–1980)—still used today—*depends* on sparking. You can find numerous YouTube videos of homemade Marx generators on the Web.

The fact that the current in an inductor can't change instantly was one of the fundamental concepts I learned to use in EE116. Another was that the voltage drop across a capacitor can't change instantly, either (again, I'll show you why this is so later). With just these two ideas, I was suddenly able to analyze all sorts of previously puzzling transient situations, and it was the *suddenness* (how appropriate for a course in *transients*!) of how I and my fellow students acquired that ability that so impressed me. To illustrate why I felt that way, here's an example of the sort of problem that EE116 treated.

In the circuit of Fig. 1, the three resistors are equal (each is R), and the two equal capacitors (C) are both uncharged. This is the situation up until the switch is closed at time $t = 0$, which suddenly connects the battery to the RC section of the circuit. The problem is to show that the current in the horizontal R first flows from right-to-left,

[5]For how arcs were used in early radio, see my book *The Science of Radio*, Springer 2001.

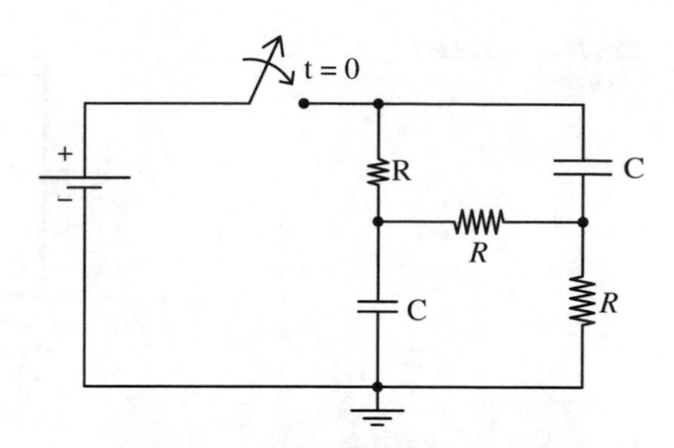

Fig. 1 A typical EE116 circuit

then gradually reduces to zero, and then reverses direction to flow left to right. Also, what is the time $t = T$ when that current goes through zero?[6] Before EE116 I didn't have the slightest idea on how to tackle such a problem, and then, suddenly, I did. *That's* why I remember EE116 with such fondness.

EE116 also cleared-up some perplexing questions that went beyond mere mathematical calculations. To illustrate what I mean by that, consider the circuit of Fig. 2, where the closing of the switch suddenly connects a *previously charged* capacitor C_1 in parallel with an uncharged capacitor, C_2. The two capacitors have different voltage drops across their terminals (just before the switch is closed, C_1's drop $\neq 0$ and C_2's drop *is* 0), voltage drops that I just told you can't change instantly. And yet, since the two capacitors are now in parallel, they *must* have the *same* voltage drop! This is, you might think, a paradox. In fact, however, we can avoid the apparent paradox *if we invoke conservation of electric charge* (the charge stored in C_1), one of the fundamental laws of physics. I'll show you how that is done, later.

Figure 3 shows another apparently paradoxical circuit that is a bit more difficult to resolve than is the capacitor circuit (but we *will* resolve it). In this new circuit, the switch has been closed for a long time, thus allowing the circuit to be in what electrical engineers call the *steady state*. Then at time $t = 0$, the switch is opened. The problem is to calculate the battery current i at just before and just after $t = 0$ (times typically written as $t = 0-$ and $t = 0+$, respectively).

For $t < 0$, the steady-state current i is the *constant* $\frac{V}{R}$ because there is no voltage drop across L_1 and, of course, there is certainly no voltage drop across the parallel L_2/switch combination.[7] So, the entire battery voltage V is across R, and Ohm's law tells us that the current in L_1 is the current in R which is $\frac{V}{R}$. This is for $t < 0$. But what is the

---

[6]The answer is $T = RC\frac{2}{3}\ln(2)$, and I'll show you later in the book how to calculate this. So, for example, if $R = 1{,}000$ ohms and $C = 0.001$ μfd, then $T = 462$ nanoseconds.

[7]The voltage-current law for an inductor L is $v_L = L\frac{di_L}{dt}$ and so, if i_L is constant, $v_L = 0$. Also, all switches in this book are modeled as perfect short-circuits when closed, and so have zero voltage drop across their terminals when closed.

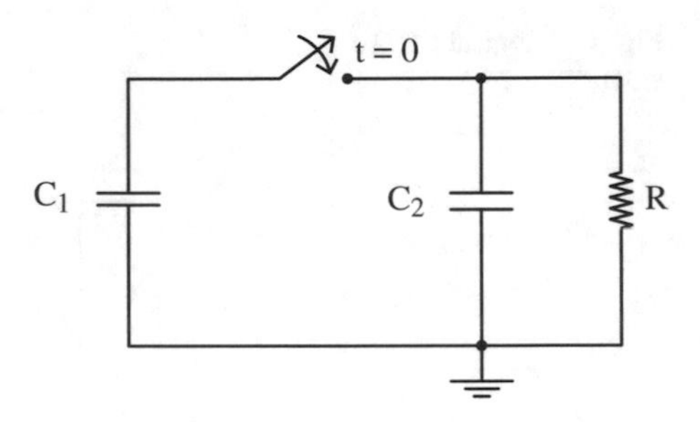

Fig. 2 A paradoxical circuit?

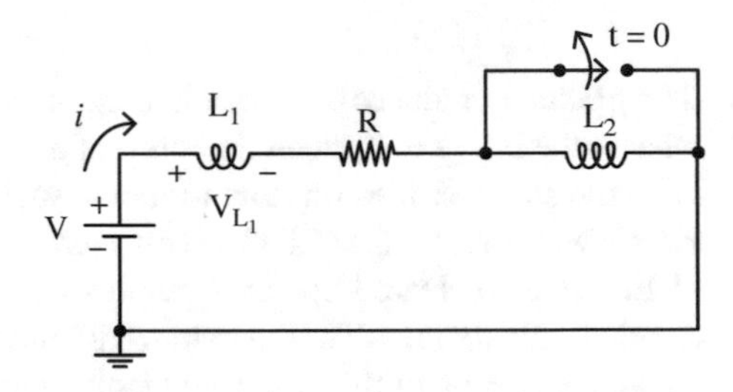

Fig. 3 Another paradoxical circuit?

current in L_2 for $t < 0$? *We don't know because, in this highly idealized circuit, that current is undefined.* You might be tempted to say it's zero because L_2 is short-circuited by the switch, but you could just as well argue that there is no current in the switch because *it's* short-circuited by L_2!

This isn't actually all that hard a puzzle to wiggle free of in "real life," however, using the following argument. Any *real* inductor and *real* switch will have *some* nonzero resistance associated with it, even if very small. That is, we can imagine Fig. 3 redrawn as Fig. 4. Resistor r_1 we can imagine absorbed into R, and so r_1 is of no impact. On the other hand, resistors r_2 and r_3 (no matter how small, just that $r_2 > 0$ and $r_3 > 0$) tell us how the current in L_1 splits between L_2 and the switch. The current in the switch is $\frac{r_2}{r_2+r_3}\frac{V}{R}$ and the current in L_2 is $\frac{r_3}{r_2+r_3}\frac{V}{R}$.

However, once we open the switch we have two inductors in series, which means they have the same current—but how can *that* be because, at $t < 0$, they generally have different currents (even in "real life") and inductor currents can't change instantly? The method of charge conservation, the method that will save the day in Fig. 2, won't work with Figs. 3 and 4; after all, *what charge*? What *will* save the day is the conservation of yet a different physical quantity, one that is a bit more subtle than is electric charge. What we'll do (at the end of Chap. 1) is derive the *conservation of total magnetic flux linking the inductors* during the switching event. When that is done, all will be resolved.

Fig. 4 A more realistic, but still paradoxical circuit

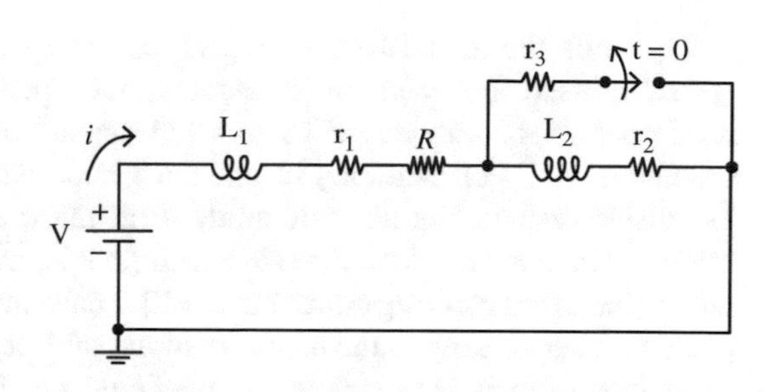

So, in addition to elementary circuit-theory,[8] that's the sort of *physics* this book will discuss. How about the *math*? Electrical circuits are mathematically described by differential equations, and so we'll be solving a lot of them in the pages that follow. If you look at older (pre-1950) electrical engineering books you'll almost invariably see that the methods used are based on something called the *Heaviside operational calculus*. This is a mathematical approach used by the famously eccentric English electrical engineer Oliver Heaviside (1850–1925), who was guided more by intuition than by formal, logical rigor. While a powerful tool in the hands of an experienced analyst who "knows how electricity works" (as did Heaviside, who early in his adult life was a professional telegraph operator), the operational method could easily lead neophytes astray.

That included many professional mathematicians who, while highly skilled in symbol manipulation, had little intuition about electrical matters. So, the operational calculus was greeted with great skepticism by many mathematicians, even though Heaviside's techniques often did succeed in answering questions about electrical circuits in situations where traditional mathematics had far less success. The result was that mathematicians continued to be suspicious of the operational calculus through the 1920s, and electrical engineers generally viewed it as something very deep, akin (almost) to Einstein's theory of general relativity that only a small, select elite could really master. Both views are romantic, fanciful myths.[9]

---

[8]The elementary circuit theory that I will be assuming really *is* elementary. I will expect, as you start this book, that you know and are comfortable with the voltage/current laws for resistors, capacitors, and inductors, with Kirchhoff's laws (in particular, loop current analysis), that an ideal battery has zero internal resistance, and that an ideal switch is a short circuit when closed and presents infinite resistance when open. I will repeat all these things again in the text as we proceed, but mostly for continuity's sake, and not because I expect you to suddenly be learning something you didn't already know. This assumed background should certainly be that of a mid-second-year major in electrical engineering or physics. As for the math, both freshman calculus and a first or second order linear differential equation should not cause panic.

[9]You can find the story of Heaviside's astonishing life (which at times seems to have been taken from a Hollywood movie) in my biography of him, *Oliver Heaviside: The Life, Work, and Times of an Electrical Genius of the Victorian Age*, The Johns Hopkins University Press 2002 (originally published by the IEEE Press in 1987). The story of the operational calculus, in particular, is told on pages 217–240. Heaviside will appear again, in the final section of this book, when we study transients in transmission lines, problems electrical engineers and physicists were confronted with in the mid-nineteenth century with the operation of the trans-Atlantic undersea cables (about which you can read in the Heaviside book, on pages 29–42).

Up until the mid-1940s electrical engineering texts dealing with transients generally used the operational calculus, and opened with words chosen to calm nervous readers who might be worried about using Heaviside's unconventional mathematics.[10] For example, in one such book we read this in the Preface: "The Heaviside method has its own subtle difficulties, especially when it is applied to circuits which are *not* 'dead' to start with [that is, when there are charged capacitors and/or inductors carrying current at $t = 0$]. I have not always found these difficulties dealt with very clearly in the literature of the subject, so I have tried to ensure that the exposition of them is as simple and methodical as I could make it."[11]

The "difficulties" of Heaviside's mathematics was specifically and pointedly addressed in an influential book by two mathematicians (using the Laplace transform years before electrical engineering educators generally adopted it), who wrote "It is doubtless because of the obscurity, not to say inadequacy, of the mathematical treatment in many of his papers that the importance of his contributions to the theory and practice of the transmission of electric signals by telegraphy and telephony was not recognized in his lifetime and that his real greatness was not then understood."[12]

One book, published 4 years before Carter's, stated that "the Heaviside operational methods [are] now widely used in [the engineering] technical literature."[13] In less than 10 years, however, that book (and all others like it[14]) was obsolete. That's because by 1949 the Laplace transform, a mathematically sound version of

[10]An important exception was the influential *graduate* level textbook (two volumes) by Murray Gardner and John Barnes, *Transients in Linear Systems: Studied by the Laplace Transformation*, John Wiley & Sons 1942. Gardner (1897–1979) was a professor of electrical engineering at MIT, and Barnes (1906–1976) was a professor of engineering at UCLA. Another exception was a book discussing the Laplace transform (using complex variables and contour integration) written by a mathematician for *advanced* engineers: R. V. Churchill, *Modern Operational Mathematics in Engineering*, McGraw-Hill 1944. Ruel Vance Churchill (1899–1987) was a professor of mathematics at the University of Michigan, who wrote several very influential books on engineering mathematics.

[11]G. W. Carter, *The Simple Calculation of Electrical Transients: An Elementary Treatment of Transient Problems in Electrical Circuits by Heaviside's Operational Methods,* Cambridge 1945. Geoffrey William Carter (1909–1989) was a British electrical engineer who based his book on lectures he gave to working engineers at an electrical equipment manufacturing facility.

[12]H. S. Carslaw and J. C. Jaeger, *Operational Methods in Applied Mathematics*, Oxford University Press 1941 (2nd edition in 1948). Horatio Scott Carslaw (1870–1954) and John Conrad Jaeger (1907–1979) were Australian professors of mathematics at, respectively, the University of Sydney and the University of Tasmania.

[13]W. B. Coulthard, *Transients in Electric Circuits Using the Heaviside Operational Calculus*, Sir Isaac Pitman & Sons 1941. William Barwise Coulthard (1893–1958) was a professor of electrical engineering at the University of British Columbia.

[14]Such books (now of only historical interest but very successful in their day) include: J. R. Carson, *Electric Circuit Theory and Operational Calculus*, McGraw-Hill 1926; V. Bush, *Operational Circuit Analysis*, Wiley & Sons 1929; H. Jeffreys, *Operational Methods in Mathematical Physics*, Cambridge University Press 1931. John Carson (1886–1940) and Vannevar Bush (1890–1974) were well-known American electrical engineers, while Harold Jeffreys (1891–1989) was an eminent British mathematician.

Heaviside's operational calculus, was available in textbook form for engineering students.[15] By the mid-1950s, the Laplace transform was firmly established as a rite of passage for electrical engineering undergraduates, and it is the central mathematical tool we'll use in this book. (When Professor Goldman's book was reprinted some years later, the words *Transformation Calculus* were dropped from the title and replaced with *Laplace Transform Theory*.)

The great attraction of the Laplace transform is the ease with which it handles circuits which are, initially, *not* "dead" (to use the term of Carter, note 11). That is, circuits in which the initial voltages and currents are other than zero. As another book told its readers in the 1930s, "A large number of Heaviside's electric circuit problems were carried out under the assumptions of initial rest and unit voltage applied at $t = 0$. These requirements are sometimes called the Heaviside condition. It should be recognized, however, that with proper manipulation, operational methods can be employed when various other circuit conditions exist."[16] With the Laplace transform, on the other hand, there is no need to think of nonzero initial conditions as requiring any special methods. The Laplace transform method of analysis is *unaltered* by, and is independent of, the initial circuit conditions.

When I took EE116 nearly 60 years ago, the instructor had to use mimeographed handouts for the class readings because there was no book available on transients *at the introductory, undergraduate level of a first course*. One notable exception might be the book *Electrical Transients* (Macmillan 1954) by G. R. Town (1905–1978) and L. A. Ware (1901–1984), who were (respectively) professors of electrical engineering at Iowa State College and the State University of Iowa. That book—which Town and Ware wrote for seniors (although they thought juniors might perhaps be able to handle much of the material, too)—does employ the Laplace transform, but specifically avoids discussing both transmission lines and the impulse function (without which much interesting transient analysis simply isn't possible), while also including analyses of then common electronic vacuum-tube circuits.[17]

---

[15]Stanford Goldman, *Transformation Calculus and Electrical Transients*, Prentice-Hall 1949. When Goldman (1907–2000), a professor of electrical engineering at Syracuse University, wrote his book it was a pioneering one for *advanced* undergraduates, but the transform itself had already been around in *mathematics* for a very long time, with the French mathematician P. S. Laplace (1749–1827) using it before 1800. However, despite being named after Laplace, Euler (see Appendix 1) had used the transform before Laplace was born (see M. A. B. Deakin, "Euler's Version of the Laplace Transform," *American Mathematical Monthly*, April 1980, pp. 264–269, for more on what Euler did).

[16]E. B. Kurtz and G. F. Corcoran, *Introduction to Electric Transients*, John Wiley & Sons 1935, p. 276. Edwin Kurtz (1894–1978) and George Corcoran (1900–1964) were professors of electrical engineering, respectively, at the State University of Iowa and the University of Maryland.

[17]Vacuum tubes *are* still used today, but mostly in specialized environments (highly radioactive areas in which the crystalline structure of solid-state devices would literally be ripped apart by atomic particle bombardment; or in high-power weather, aircraft, and missile-tracking radars; or in circuits subject to nuclear explosion electromagnetic pulse—EMP—attack, such as electric power-grid electronics), but you'd have to look hard to find a vacuum tube in any everyday consumer product (and certainly not in modern radio and television receivers, gadgets in which the soft glow of red/yellow-hot filaments was once the very signature of electronic circuit mystery).

I will say a lot more about the impulse function later in the book, but for now let me just point out that even after its popularization among physicists in the late 1920s by the great English mathematical quantum physicist Paul Dirac (1902–1984)[18]—it is often called the *Dirac delta function*—it was still viewed with not just a little suspicion by both mathematicians and engineers until the early 1950s.[19] For that reason, perhaps, Town and Ware avoided its use. Nonetheless, their book was, in my opinion, a very good one for its time, but it would be considered dated for use in a modern, first course. Finally, in addition to the book by Town and Ware, there is one other book I want to mention because it was so close to my personal experience at Stanford.

Hugh H. Skilling (1905–1990) was a member of the electrical engineering faculty at Stanford for decades and, by the time I arrived there, he was the well-known author of electrical engineering textbooks in circuit theory, transmission lines, and electromagnetic theory. Indeed, at one time or another, during my 4 years at Stanford, I took classes using those books and they were excellent treatments. A puzzle in this, however, is that in 1937 Skilling also wrote another book called *Transient Electric Currents* (McGraw-Hill), which came out in a second edition in 1952. The reason given for the new edition was that the use of Heaviside's operational calculus in the first edition needed to be replaced with the Laplace transform. That was, of course, well and good, as I mentioned earlier—so why wasn't the new 1952 edition of Skilling's book used in my EE116 course? It was obviously available when I took EE116 8 years later but, nonetheless, was passed over. Why? Alas, it's too late now to ask my instructor from nearly 60 years ago—Laurence A. Manning (1923–2015)[20]—but here's my guess.

Through the little-picture eyes of a 20-year-old student, I thought Professor Manning was writing an introductory book on electrical transients, one built around the fundamental ideas of how current and voltage behave in suddenly switched circuits built from resistors, capacitors, and inductors. I thought it was going to be a book making use of the so-called *singular* impulse function and, perhaps, too, an elementary treatment of the Laplace transform would be part of the book. Well, I was wrong about all that.

But I didn't realize that until many years later, when I finally took a look at the book he *did* write and publish 4 years after I had left Stanford: *Electrical Circuits* (McGraw-Hill 1966). This is a very broad (over 550 pages long) work that discusses the steady-state AC behavior of circuits, as well as nonelectrical (that is,

[18]Dirac, who had a PhD in mathematics and was the Lucasian Professor of *Mathematics* at Cambridge University (a position held, centuries earlier, by Isaac Newton), received a share of the 1933 Nobel Prize in *physics*. Before all that, however, Dirac had received first-class honors at the University of Bristol as a 1921 graduate in *electrical engineering*.

[19]This suspicion was finally removed with the publication by the French mathematician Laurent Schwartz (1915–2002) of his *Theory of Distributions*, for which he received the 1950 Fields Medal, the so-called "Nobel Prize of mathematics."

[20]Professor Manning literally spent his entire life at Stanford, having been born there, on the campus where his father was a professor of mathematics.

mechanically analogous) systems. There *are* several chapters dealing with transients, yes, but lots of other stuff, too, and that other material accounts for the majority of those 550 pages. The development of the Laplace transform is, for example, taken up to the level of the inverse transform contour integral evaluated in the complex plane.

In the big-picture eyes that I think I have now, Professor Manning's idea of the book he was writing was far more extensive than "just" one on transients for EE116. As he wrote in his Preface, "The earlier chapters have been used with engineering students of all branches at the sophomore level,"[21] while "The later chapters continue the development of circuit concepts through [the] junior-year [EE116, for example]." The more advanced contour integration stuff, in support of the Laplace transform, was aimed at seniors and first-year graduate students. All those mimeographed handouts I remember were simply for individual *chapters* in his eventual book.

Skilling's book was simply too narrow, I think, for Professor Manning (in particular, its lack of discussion on impulse functions), and that's why he passed it by for use in EE116—and, of course, he wanted to "student test" the transient chapter material he was writing for his own book. But, take Skilling's transient book, add Manning's impulse function material, along with a non-contour integration presentation of the Laplace transform, *all the while keeping it short (under 200 pages)*, then that *would* have been a neat *little* book for EE116. I've written this book as that missing *little* book, the book I wish had been available all those years ago.

So, with that goal in mind, this book is aimed at mid to end-of-year sophomore or beginning junior-year electrical engineering students. While it has been written under the assumption that readers are encountering *transient* electrical analysis for the first time, the mathematical and physical theory is not "watered-down." That is, the analysis of both lumped and continuous (transmission line) parameter circuits is performed with the use of differential equations (both ordinary and partial) in the time domain and in the Laplace transform domain. The transform is fully developed (*short of invoking complex variable analysis*) in the book for readers who are *not* assumed to have seen the transform before.[22] The use of singular time functions (the unit step and impulse) is addressed and illustrated through detailed examples.

---

[21] I think Professor Manning is referring here to non-electrical engineering students (civil and mechanical, mostly) who needed an electrical engineering elective, and so had selected the sophomore circuits course that the Stanford EE Department offered to non-majors (a common practice at all engineering schools).

[22] The way complex variables usually come into play in transient analysis is during the inversion of a Laplace transform back to a time function. This typical way of encountering transform theory has resulted in the common belief that it is *necessarily* the case that transform inversion *must* be done via contour integration in the complex plane: see C. L. Bohn and R. W. Flynn, "Real Variable Inversion of Laplace Transforms: An Application in Plasma Physics," *American Journal of Physics*, December 1978, pp. 1250–1254. In this book, all transform operations will be carried out as *real* operations on *real* functions of a *real* variable, making all that we do here mathematically completely accessible to lower-division undergraduates.

One feature of this book, that the authors of yesteryear could only have thought of as science fiction, or even as being sheer fantasy, is the near-instantaneous electronic evaluation of complicated mathematics, like solving numerous simultaneous equations with all the coefficients having ten (or more) decimal digits. Even after the Heaviside operational calculus was replaced by the Laplace transform, there often is still much tedious algebra to wade through for any circuit using more than a handful of components. With a modern scientific computing language, however, much of the horrible symbol-pushing and slide-rule gymnastics of the mid-twentieth century has been replaced at the start of the twenty-first century with the typing of a single command. In this book I'll show you how to do that algebra, but often one can avoid the worst of the miserable, grubby arithmetic with the aid of computer software (or, at least, one can check the accuracy of the brain-mushing hand-arithmetic). In this book I use *MATLAB*, a language now commonly taught worldwide to electrical engineering undergraduates, often in their freshman year. Its use here will mostly be invisible to you—I use it to generate all the plots in the book, for the inversion of matrices, and to do the checking of some particularly messy Laplace transforms. This last item doesn't happen much, but it *does* ease concern over stupid mistakes caused by one's eyes glazing over at all the number-crunching.

The appearance of paradoxical circuit situations, often ignored in many textbooks (because they are, perhaps, considered "too advanced" or "confusing" to explain to undergraduates in a first course) is fully embraced as an opportunity to challenge readers. In addition, historical commentary is included throughout the book, to combat the common assumption among undergraduates that all the stuff they read in engineering textbooks was found engraved on Biblical stones, rather than painfully discovered by people of genius who often first went down a *lot* of false rabbit holes before they found the right one.

Durham, NH, USA Paul J. Nahin

Acknowledgments

An author, alone, does not make a book. There are other people involved, too, providing crucial support, and my grateful thanks goes out to all of them. This book found initial traction at Springer with the strong support of my editor, Dr. Sam Harrison, and later I benefited from the aid of editorial assistant Sanaa Ali-Virani. My former colleague at Harvey Mudd College in Claremont, California, professor emeritus of engineering Dr. John Molinder, read the entire book; made a number of most helpful suggestions for improvement; and graciously agreed to contribute the Foreword. The many hundreds of students I have had over more than 30 years of college teaching have had enormous influence on my views of the material in this book.

Special thanks are also due to the ever-pleasant staff of *Me & Ollie's Bakery, Bread and Café* shop on Water Street in Exeter, New Hampshire. As I sat, almost daily for many months, in my cozy little nook by a window, surrounded by happily chattering Phillips Exeter Academy high school students from just up the street (all of whom carefully avoided eye contact with the strange old guy mysteriously scribbling away on papers scattered all over the table), the electrical mathematics and computer codes seemed to just roll off my pen with ease.

Finally, I thank my wife Patricia Ann who, for 55 years, has put up with manuscript drafts and reference books scattered all over her home. With only minor grumbling (well, maybe not *always* so minor) she has allowed my inner-slob free reign. Perhaps she has simply given up trying to change me, but I prefer to think it's because she loves me. I know I love her.

University of New Hampshire
Durham and Exeter, NH, USA
January 2018

Paul J. Nahin

Contents

1 Basic Circuit Concepts 1
1.1 The Hardware of Circuits 1
1.2 The Physics of Circuits 4
1.3 Power, Energy, and Paradoxes 7
1.4 A Mathematical Illustration 12
1.5 Puzzle Solution 18
1.6 Magnetic Coupling, Part 1 21

2 Transients in the Time Domain 29
2.1 Sometimes You Don't Need a Lot of Math 29
2.2 An Interesting Switch-Current Calculation 31
2.3 Suppressing a Switching Arc 37
2.4 Magnetic Coupling, Part 2 41

3 The Laplace Transform 51
3.1 The Transform, and Why It's Useful 51
3.2 The Step, Exponential, and Sinusoid Functions of Time 55
3.3 Two Examples of the Transform in Action 61
3.4 Powers of Time 67
3.5 Impulse Functions 75
3.6 The Problem of the Reversing Current 79
3.7 An Example of the Power of the Modern Electronic Computer . . . 83
3.8 Puzzle Solution 88
3.9 The Error Function and the Convolution Theorem 93

4 Transients in the Transform Domain 105
4.1 Voltage Surge on a Power Line 105
4.2 Two Hard Problems from Yesteryear 113
4.3 Gas-Tube Oscillators 120
4.4 A Constant Current Generator 125

5 Transmission Lines 133
5.1 The Partial Differential Equations of Transmission Lines 133
5.2 Solving the Telegraphy Equations 137
5.3 The Atlantic Cable 141
5.4 The Distortionless Transmission Line 145
5.5 The General, Infinite Transmission Line 147
5.6 Transmission Lines of Finite Length 154

Appendix 1: Euler's Identity 161

Appendix 2: Heaviside's Distortionless Transmission Line Condition 167

Appendix 3: How to Solve for the Step Response of the Atlantic Cable Diffusion Equation *Without* the Laplace Transform 171

Appendix 4: A Short Table of Laplace Transforms and Theorems . . . 185

Index 187

About the Author

Paul Nahin was born in California and did all of his schooling there (Brea-Olinda High 1958, Stanford BS 1962, Caltech MS 1963, and – as a Howard Hughes Staff Doctoral Fellow – UC/Irvine PhD 1972, with all degrees in electrical engineering). He has taught at Harvey Mudd College, the Naval Postgraduate School, and the universities of New Hampshire (where he is now Emeritus Professor of Electrical Engineering) and Virginia.

Prof. Nahin has published a couple of dozen short science fiction stories in *Analog*, *Omni*, and *Twilight Zone* magazines, and has written 19 books on mathematics and physics for scientifically minded and popular audiences alike. He has given invited talks on mathematics at Bowdoin College, the Claremont Graduate School, the University of Tennessee and Caltech, has appeared on National Public Radio's "Science Friday" show (discussing time travel) as well as on New Hampshire Public Radio's "The Front Porch" show (discussing imaginary numbers), and advised Boston's WGBH Public Television's "Nova" program on the script for their time travel episode. He gave the invited Sampson Lectures for 2011 in Mathematics at Bates College (Lewiston, Maine). He received the 2017 Chandler Davis prize for Excellence in Expository Writing in Mathematics.

Also by Paul J. Nahin

Oliver Heaviside: *The Life, Work, and Times of an Electrical Genius of the Victorian Age*
Time Machines: *Time Travel in Physics, Metaphysics, and Science Fiction*
The Science of Radio: *With MATLAB® and Electronics Workbench® Demonstrations*
An Imaginary Tale: *The Story of √-1*
Duelling Idiots: *And Other Probability Puzzlers*
When Least Is Best: *How Mathematicians Discovered Many Clever Ways to Make Things as Small (or as Large) as Possible*
Dr. Euler's Fabulous Formula: *Cures Many Mathematical Ills*
Chases and Escapes: *The Mathematics of Pursuit and Evasion*
Digital Dice: *Computational Solutions to Practical Probability Problems*
Mrs. Perkins's Electric Quilt: *And Other Intriguing Stories of Mathematical Physics*
Time Travel: *A Writer's Guide to the Real Science of Plausible Time Travel*
Number-Crunching: *Taming Unruly Computational Problems from Mathematical Physics to Science Fiction*
The Logician and the Engineer: *How George Boole and Claude Shannon Created the Information Age*
Will You Be Alive Ten Years from Now?: *And Numerous Other Curious Questions in Probability*
Holy Sci-Fi!: *Where Science Fiction and Religion Intersect*
Inside Interesting Integrals (with an introduction to contour integration)
In Praise of Simple Physics: *The Science and Mathematics Behind Everyday Questions*
Time Machine Tales: *The Science Fiction Adventures and Philosophical Puzzles of Time Travel*

Chapter 1
Basic Circuit Concepts

1.1 The Hardware of Circuits

There are three fundamental components commonly used in electrical/electronic circuitry: resistors, capacitors, and inductors (although this last component will get some qualifying remarks in just a bit). Another component commonly encountered is the transformer and it will get some discussion, too, later. All of these components are passive. That is, they do not generate electrical energy, but either dissipate energy as heat (resistors) or temporarily store energy in an electric field (capacitors) or in a magnetic field (inductors). Transformers involve magnetic fields, as do inductors, but do *not* store energy. We'll return to transformers later in the book. The first three components have two-terminals (the transformer in its simplest form has four), as shown in Fig. 1.1.

There are, of course, other more complex, multi-terminal components used in electrical/electronic circuits (most obviously, transistors), as well as such things as constant voltage, and constant current, *sources*,[1] but for our introductory treatment of transients, these three will be where we'll concentrate our attention. We can formally define each of the three passive, two-terminal components by the relationship that connects the current (i) through them to the voltage drop (v) across them. If we denote the values of these components by R (ohms), C (farads), and L (henrys), and if v and i have the unit of volts and amperes, named after the Italian scientist Alessandro Volta (1745–1827) and the French mathematical physicist André Marie Ampere (1775–1836), respectively, and if time (t) is in units of seconds, then the

[1] Sources are *not* passive, as they are the origins of energy in an electrical circuit. A constant voltage source maintains a fixed voltage drop across its terminals, independent of the current in it (think of the common battery). Constant current sources maintain a fixed current in themselves, independent of the voltage drop across their terminals, and are *not* something you can buy in the local drugstore like a battery. You have to construct them. In Chap. 4 I'll show you how to make a *theoretically perfect* (after the transients have died away) a-c *constant current* generator out of just inductors, capacitors, and a sinusoidal *voltage* source.

P. J. Nahin, *Transients for Electrical Engineers*,
https://doi.org/10.1007/978-3-319-77598-2_1

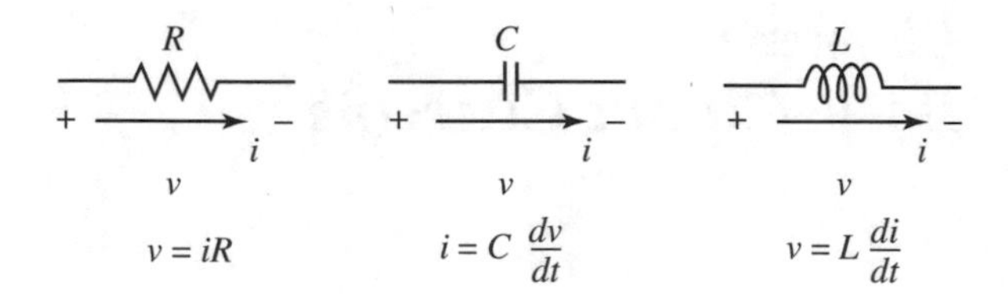

Fig. 1.1 The three standard, passive, two-terminal components

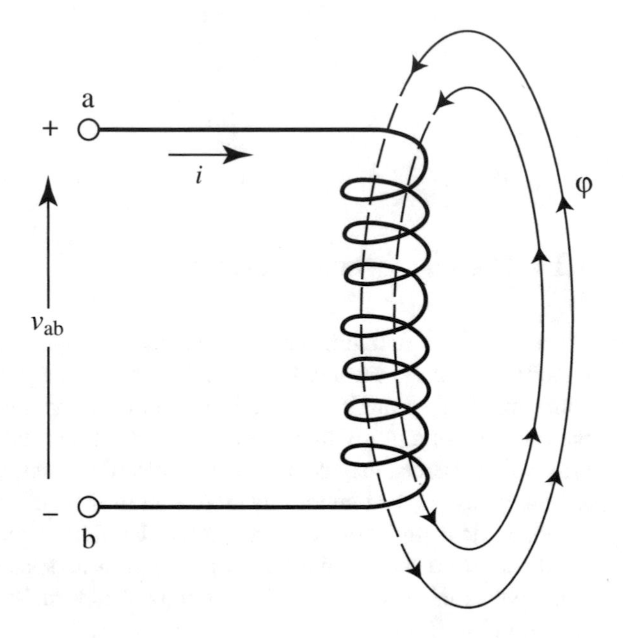

Fig. 1.2 A current-carrying coil, with magnetic flux ϕ

mathematical definitions of the components are as shown in Fig. 1.1. Of course, $v = iR$ is the famous Ohm's law, named after the German Georg Ohm (1787–1854). The other two relationships don't have commonly used names, but the units of capacitance and inductance are named, respectively, after the English experimenter Michael Faraday (1791–1867) and the American physicist Joseph Henry (1799–1878). As a general guide, 1 ohm is a small resistance, 1 farad is very large capacitance, and 1 henry is a fairly large inductance. The possible ranges on voltages and currents is enormous, ranging from micro-volts/micro-amps to mega-volts/mega-amps.

The current-voltage laws of the resistor and the capacitor are sufficient in themselves for what we'll do in this book (that is, we don't need to delve more deeply into 'how they work'), but for the inductor we do need to say just a bit more. So, imagine that a coil of wire, with n turns, is carrying a current $i(t)$, as shown in Fig. 1.2. The current creates a magnetic field of closed (no ends) flux lines that *encircle* or *thread through* the turns of the coil. (More on flux, later in this chapter.) Ampere's law says that the flux produced by each turn of the coil is proportional to i, that is, the contribution by each turn to the total flux ϕ is Ki, where K is some constant depending on the size of the coil and the nature of the matter inside the coil.

Since the flux contributions add, then the total flux produced by the n turns is $\phi = nKi$. Now, from Faraday's law of induction, a *change* in the flux through a turn of the coil produces a potential difference in *each* turn of the coil of magnitude $d\phi/dt$. Since there are n turns in series, then the total potential difference that appears across the ends of the coil has magnitude

$$v_{ab} = n\frac{d\phi}{dt} = n\frac{d(nKi)}{dt} = Kn^2\frac{di}{dt} = L\frac{di}{dt} \tag{1.1}$$

where $L = Kn^2$ is the so-called *self-inductance* of an n-turn coil (notice that L varies as the *square* of the number of turns). If we define *flux linkage* as the product of the number of turns in a coil and the flux linking (passing through) each turn, that is as

$$\phi_L = n\phi,$$

then (1.1) can be written as

$$v_{ab} = \frac{d(n\phi)}{dt} = \frac{d\phi_L}{dt} = \frac{d(Li)}{dt}.$$

Thus, to within a constant (which we'll take as zero since $\phi_L(i = 0) = 0$), we have

$$\phi_L = Li. \tag{1.2}$$

This result, as you'll see at the end of this chapter, is the key to resolving the paradox mentioned in the Preface in connection with Fig. 3.

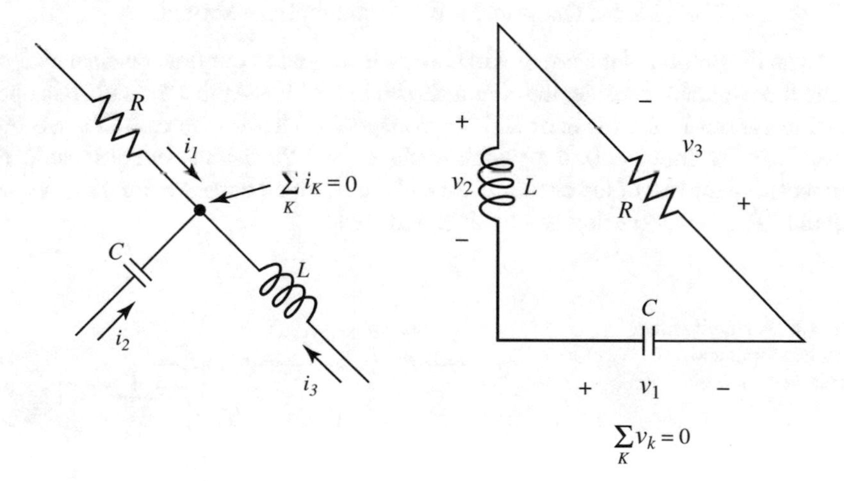

Fig. 1.3 Kirchhoff's two circuit laws

1.2 The Physics of Circuits

In all of our analyses, we will routinely use two 'laws' (dating from 1845) named after the German physicist Gustav Robert Kirchhoff (1824–1887). These two laws, illustrated in Fig. 1.3, are in fact actually the fundamental physical laws of the conservation of energy and the conservation of electric charge.

Kirchhoff's voltage law The sum of the *voltage (or electric potential) drops* around any closed path (loop) in a circuit is zero. Voltage is defined to be energy per unit charge, and the voltage *drop* is the energy expended in transporting a unit charge through the electric field that exists inside the component. The law, then, says that the net energy change for a unit charge transported around a closed path is zero. If it were not zero, then we could repeatedly transport charge around the closed path in the direction in which the net energy change is positive and so become rich selling the energy gained to the local power company. Conservation of energy, however, says we can't do that. (Since the sum of the drops is zero, then one can also set the sum of the voltage *rises* around any closed loop to zero.)

Kirchhoff's current law The sum of the currents into any point in a circuit is zero. This says that if we construct a tiny, closed surface around any point in a circuit then the charge enclosed by that surface remains constant. That is, whatever charge is transported into the enclosed volume by one current is transported out of the volume by other currents; current *is* the motion of electric charge. Mathematically, the current i at any point in a circuit is defined to be the rate at which charge is moving through that point, that is, $i = dQ/dt$. Q is measured in coulombs — named after the French physicist Charles Coulomb (1736–1806) — where the charge on an electron is 1.6×10^{-19} coulombs. One ampere is one coulomb per second.

As an illustration of the use of Kirchhoff's laws, and as our first transient analysis in the time domain, consider the circuit shown in Fig. 1.4. At first the switch is open, and the current in the inductor and the voltage drop across the capacitor are both zero. Then, at time $t = 0$, the switch is closed and the 1-volt battery is suddenly connected to the rest of the circuit. If we call the resulting battery current $i(t)$, we can calculate $i(t)$ for $t > 0$ using Kirchhoff's two laws.

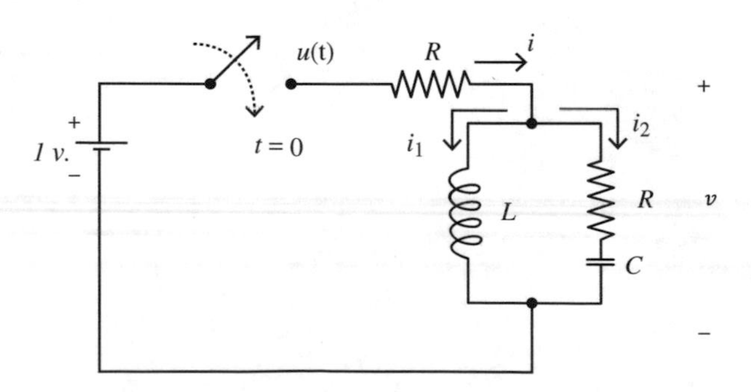

Fig. 1.4 A circuit with a switched input and a transient response

Using the notation of Fig. 1.4, and the two laws, we can write the following set of equations:

$$i = i_1 + i_2, \tag{1.3}$$

$$u = iR + v, \tag{1.4}$$

$$v = L\frac{di_1}{dt}, \tag{1.5}$$

$$v = i_2R + \frac{1}{C}\int_0^t i_2(x)dx. \tag{1.6}$$

These four equations completely describe the behavior of $i(t)$ for all $t > 0$.[2] If we differentiate[3] (1.6) with respect to time, then we can also write

$$\frac{dv}{dt} = R\frac{di_2}{dt} + \frac{1}{C}i_2. \tag{1.7}$$

We can manipulate and combine these equations to eliminate the variables i_1 and i_2, to arrive at the following second-order, linear differential equation relating $u(t)$, the applied voltage, to the resulting current $i(t)$.

$$\frac{d^2u}{dt^2} + \left(\frac{R}{L}\right)\frac{du}{dt} + \left(\frac{1}{LC}\right)u = 2R\frac{d^2i}{dt^2} + \left(\frac{R^2}{L} + \frac{1}{C}\right)\frac{di}{dt} + \left(\frac{R}{LC}\right)i. \tag{1.8}$$

When I say "we can manipulate and combine" the equations of the circuit in Fig. 1.4, I don't mean doing that is necessarily *easy* to do, at least not as the equations stand (in the time domain). You should *try* to confirm (1.8) for yourself, and later you'll see (and *greatly* appreciate!) just how much easier it will be when we get to the Laplace transform.

[2] The last term in (1.6) comes from integrating the equation $i_2 = C\frac{dv_C}{dt}$, where v_C is the voltage drop *across the capacitor*. If V_0 is the voltage drop across the capacitor *at* time $t = 0$ (the so-called *initial voltage*), then we have the voltage drop across the capacitor for any time $t \geq 0$ as $v_C(t) = \frac{1}{C}\int_0^t i_2(x)dx + V_0$ where x is a so-called *dummy variable of integration*. For the circuit in Fig. 1.4, we are given that $V_0 = 0$.

[3] To differentiate an integral, all electrical engineers should know *Leibniz's formula*, named after the German mathematician Gottfried Wilhelm Leibniz (1646–1716). I won't derive it here, but you can find a proof in any good book in advanced calculus: If $g(y) = \int_{v(y)}^{u(y)} f(x, y)dx$, then $\frac{dg}{dy} = \int_{v(y)}^{u(y)} \frac{\partial f}{\partial y}dx + f\{u(y), y\}\frac{du}{dy} - f\{v(y), y\}\frac{dv}{dy}$, where $\frac{\partial f}{\partial y}$ is the *partial* derivative of $f(x, y)$ with respect to y. For the special case where $u(y)$ and $v(y)$ are *constants*, then the last two terms are each zero and the derivative of the integral is the integral of the derivative (this is called *differentiating under the integral sign*).We'll use this when we get to the Laplace transform.

Since $u(t) = 1$ for $t > 0$, we have

$$\frac{du}{dt} = \frac{d^2u}{dt^2} = 0, \quad t > 0$$

and so, for $t > 0$, the differential equation for $i(t)$ reduces to

$$2R\frac{d^2i}{dt^2} + \left(\frac{R^2}{L} + \frac{1}{C}\right)\frac{di}{dt} + \left(\frac{R}{LC}\right)i = \frac{1}{LC}. \tag{1.9}$$

We can solve (1.9) using the standard technique of writing $i(t)$ as the sum (because the differential equation is linear) of the solutions for the *homogeneous* case (set the right-hand-side of (1.9) equal to zero), and the 'obvious' solution of i equal to the *constant* $\frac{1}{R}$. (Mathematicians call this the *particular* solution.) The particular solution is 'obvious' because, with i equal to a constant, we clearly have

$$\frac{di}{dt} = \frac{d^2i}{dt^2} = 0$$

and (1.9) reduces to

$$\left(\frac{R}{LC}\right)i = \frac{1}{LC}.$$

That is, (1.9) reduces to $i = \dfrac{1}{R}$.

To solve the homogeneous case, we have to do a bit more work. Again, the standard technique is to *assume* a solution of the form

$$i(t) = Ie^{st}$$

where I and s are both constants (perhaps complex). This immensely clever idea (the origin of which is buried in the history of mathematics) works because every time derivative of e^{st} is simply a power of s multiplied by e^{st}, and so all the e^{st} factors will cancel away. To see this, substitute the *assumed* solution into the homogeneous differential equation to get

$$2Rs^2Ie^{st} + \left(\frac{R^2}{L} + \frac{1}{C}\right)sIe^{st} + \left(\frac{R}{LC}\right)Ie^{st} = 0.$$

Dividing through by the common factor of Ie^{st}, we are left with simply a quadratic, *algebraic* (*not* differential) equation in s:

$$s^2 + \frac{R^2C + L}{2RLC}s + \frac{1}{2LC} = 0. \tag{1.10}$$

Let's call the two roots (perhaps complex-valued) of this quadratic s_1 and s_2, the values of which we see are completely determined by the circuit component values.

Then, the general solution for the current $i(t)$ is the sum of the particular solution and the two homogeneous solutions:

$$i(t) = \frac{1}{R} + I_1 e^{s_1 t} + I_2 e^{s_2 t} \tag{1.11}$$

where I_1 and I_2 are constants yet to be determined. To find them, we need to step away from (1.11) for just a bit, and discover (at last) the way currents in inductors, and voltage drops across capacitors, can (or cannot) change in zero time (that is, instantaneously).

1.3 Power, Energy, and Paradoxes

The instantaneous power $p(t)$ is the *rate* at which energy is delivered to a component, and is given by

$$p(t) = v(t)i(t) \tag{1.12}$$

where $p(t)$ has the units of watts (1 watt = 1 joule/second), v (the voltage drop across the component) is in volts, and i (the current in the component) is in amperes.[4] To see that this is dimensionally correct, first note that power is energy per unit time. Then, recall that voltage is energy per unit charge, and that current is charge per unit time. Thus, the product vi has units (energy/charge) times (charge/time) = energy/time, the unit of power. If we integrate power over an interval of time, the result is the total energy (W) delivered to that component during that time interval.

For example, for a resistor we have $v = iR$ and so

$$p = vi = (iR)i = i^2 R \tag{1.13}$$

or, in the time interval 0 to T, the total energy delivered to the resistor is

$$W = \int_0^T p(t)dt = \int_0^T (i^2)Rdt = R\int_0^T i^2(t)dt. \tag{1.14}$$

Since the integrand is always nonnegative (you can't get anything negative by *squaring* a real quantity) we conclude, independent of the time behavior of the current, that $W > 0$ if $i(t) \neq 0$. The electrical energy delivered to a resistor is totally converted to heat energy, that is, the temperature of a resistor carrying current increases.

[4]The units of watts and joules are named after, respectively, the Scottish engineer James Watt (1736–1819) and the English physicist James Joule (1818–1889). Like the ampere, volt, and ohm, watts and joules are units in the MKS (meter/kilogram/second) *metric* system. To give you an idea of what a joule is, burning a gallon of gasoline releases about 100 mega-joules of chemical energy.

For inductors and capacitors, however, the situation is remarkably different. For an inductor, for example,

$$p = vi = \left(L\frac{di}{dt}\right)i = \frac{1}{2}L\frac{d(i^2)}{dt} \tag{1.15}$$

and so

$$W = \frac{1}{2}L\int_0^T \frac{d(i^2)}{dt}dt = \frac{1}{2}L\int_0^T d(i^2). \tag{1.16}$$

The limits on the last integral are in units of *time*, while the variable of integration is i. It is thus perhaps clearer to write W as

$$W = \frac{1}{2}L\int_{i^2(0)}^{i^2(T)} d(i^2) = \frac{1}{2}L[i^2(T) - i^2(0)] \tag{1.17}$$

where $i(0)$ and $i(T)$ are the inductor *currents* at times $t = 0$ and $t = T$, respectively. Thus, $i(t)$ can have *any physically possible behavior* over the time interval from 0 to T and yet, if $i(0) = i(T)$ then $W = 0$. What happens, physically, is that as $i(t)$ varies from its initial value at time $t = 0$, energy is *stored* in the magnetic field around the inductor and then, as the current returns to its initial value at time $t = T$, the *stored* energy is returned to the circuit (that is, to the original source of the energy) as the field 'collapses.'[5] Inductors do not dissipate electrical energy by converting it into heat energy and so, unlike resistors, ideal inductors don't get warm when carrying current.

We can do a similar analysis for capacitors. The power to a capacitor is

$$p = vi = vC\frac{dv}{dt} = \frac{1}{2}C\frac{d(v^2)}{dt} \tag{1.18}$$

and so the energy over the time interval 0 to T

$$W = \frac{1}{2}C\int_0^T \frac{d(v^2)}{dt}dt = \frac{1}{2}C\int_0^T d(v^2). \tag{1.19}$$

Again, the limits on the last integral are in units of *time*, while the variable of integration is v^2. It is thus perhaps clearer to write W as

[5]This is all very picturesque language, but you shouldn't confuse it with actual knowledge. Just *how* energy is stored in what appears to be empty space, in an invisible magnetic field, is a very mysterious thing, and I think any honest physicist and electrical engineer would admit that. In the same way, while the imagery of magnetic flux lines is 'suggested' by looking at how iron filings line-up on a piece of paper held over a magnet, I think anybody who believes simply saying the words *magnetic flux line* represents actual understanding is fooling themselves.

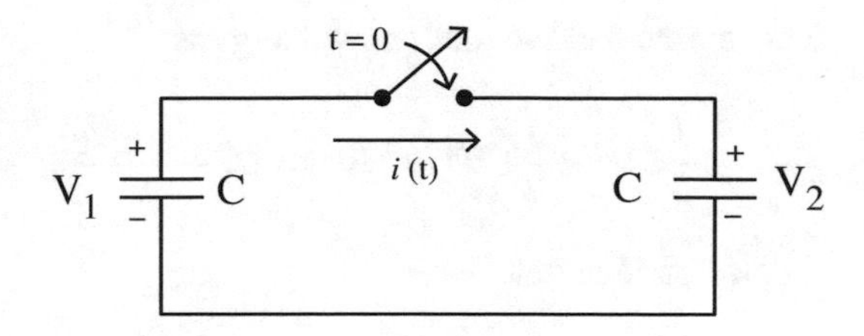

Fig. 1.5 A famous two-capacitor puzzle

$$W = \frac{1}{2}C\int_{v^2(0)}^{v^2(T)} d(v^2) = \frac{1}{2}C\left[v^2(T) - v^2(0)\right]. \tag{1.20}$$

This tells is that $W = 0$ *if* $v(0) = v(T)$, and the energy (now stored in an *electric* field, and see note 5 again) Is returned from *temporary* storage to the circuit.

These power and energy concepts may appear to be quite elementary to you, but if you consider the following two little puzzles I think you might rethink that impression. First, suppose we have two equal capacitors that can be connected together by a switch, as shown in Fig. 1.5. Before the switch is closed at time $t = 0$, C_1 is charged to V_1 volts and C_2 is charged to V_2 volts, where $V_1 \neq V_2$. Thus, for $t < 0$ the total charge is

$$CV_1 + CV_2 = C(V_1 + V_2)$$

and the total, initial energy is

$$\frac{1}{2}CV_1^2 + \frac{1}{2}CV_2^2 = \frac{1}{2}C\left(V_1^2 + V_2^2\right) = W_i.$$

When the switch is closed, the charge (which is conserved) will redistribute itself (via the current $i(t)$) between the two capacitors so that the capacitors will have the same voltage drop V. So, for $t > 0$, the total charge is

$$CV + CV = 2CV$$

and, because charge is conserved, we have

$$2CV = C(V_1 + V_2)$$

or,

$$V = \frac{V_1 + V_2}{2}.$$

This means that the total, final energy is

$$\frac{1}{2}CV^2+\frac{1}{2}CV^2=CV^2=C\left(\frac{V_1+V_2}{2}\right)^2=\frac{1}{2}C\frac{(V_1+V_2)^2}{2}=W_f.$$

Now, notice that

$$W_i-W_f=\frac{1}{2}C(V_1^2+V_2^2)-\frac{1}{2}C\frac{(V_1+V_2)^2}{2}=\frac{1}{2}C\left[V_1^2+V_2^2-\frac{V_1^2+2V_1V_2+V_2^2}{2}\right]$$
$$\frac{1}{2}C\left[\frac{2V_1^2+2V_2^2-V_1^2-2V_1V_2-V_2^2}{2}\right]=\frac{1}{4}C\left[V_1^2-2V_1V_2+V_2^2\right]$$
$$=\frac{1}{4}C(V_1-V_2)^2>0$$

because *any* real number squared is never negative (and is zero only if $V_1=V_2$). Thus, $W_i\neq W_f$ and energy has *not* been conserved. The final energy is *less* than the initial energy. Where did the missing energy go? Think about all this as you continue to read, and I'll show you a way out of the puzzle in Sect. 1.5.

For a second, even more puzzling quandary, suppose we have a resistor *R* with current *i*(*t*) in it. Then, as before in (1.14),

$$W=\int_{-\infty}^{\infty}p(t)dt=R\int_{-\infty}^{\infty}i^2(t)dt$$

as the total energy dissipated as heat by the resistor over *all* time (notice the limits of integration). Also, as current is the time derivative of electric charge, we have

$$Q=\int_{-\infty}^{\infty}i(t)dt$$

as the total charge that passes through the resistor. Consider now the following *specific* *i*(*t*): for *c* some positive finite *constant*, let

$$i(t)=\begin{cases}0, & t<0\\ c^{-4/5}, 0<t<c & .\\ 0, & t>c\end{cases}$$

That is, *i*(*t*) is a *finite*-valued pulse of current that is non-zero over a *finite* period of time. The total charge transported through the resistor is

$$Q=\int_0^c c^{-4/5}dt=c^{-4/5}c=c^{1/5}.$$

If we pick the constant *c* to be ever smaller, that is, if we let $c\to 0$, then the pulse-like current obviously does something a bit odd — it becomes ever briefer in duration but ever larger in amplitude in such a way that $\lim_{c\to 0}Q=0$. That is, even though the amplitude of the current pulse blows-up, the pulse duration becomes shorter 'even faster' so that the total charge transported through the resistor goes to zero. Now, what's the puzzle in all this?

We have, over the duration of the current pulse,

$$i^2(t) = c^{-8/5}$$

and so

$$W = R\int_0^c c^{-8/5}dt = Rc^{-8/5}c = Rc^{-3/5}.$$

So,

$$\lim_{c\to 0} W = \infty$$

which means the resistor will instantly vaporize because all that infinite energy is delivered in zero time. But how can *that* be, as we just showed in the limit of $c \to 0$ there is *no charge* transported through the resistor? Think about this as you continue to read and, at the end of Chap. 3, after we've developed the Laplace transform, I'll show you a way out of the fog.

Now, to finish this section, let's look (finally!) at how currents in inductors and voltage drops across capacitors can (or cannot) change. The instantaneous power in each is given, respectively, by

$$p_L = vi = L\frac{di}{dt}i$$

and

$$p_C = vi = vC\frac{dv}{dt}.$$

If the current in an inductor, or the voltage drop across a capacitor, could change instantaneously, then we would have

$$\frac{di}{dt} = \infty, \frac{dv}{dt} = \infty$$

which give infinite power. But electrical engineers and physicists reject the idea of any physical quantity, in *any* circuit we could actually build, having an infinite value. So, inductor currents and capacitor voltage drops *cannot* change instantaneously.

As an immediate (and important) corollary to this (since the magnetic flux ϕ of an inductor is directly proportional to the current in the inductor) is that if an inductor current cannot change instantly then neither can the flux: inductor magnetic flux is a *continuous* function of time. Note, however, that since the power in a resistor does not include a time derivative, both the voltage drop across, and the current in, a resistor *can* change instantly.

1.4 A Mathematical Illustration

With the final result of the previous section, we can now return to (1.11) and solve for I_1 and I_2. Repeating (1.11) as (1.21),

$$i(t) = \frac{1}{R} + I_1 e^{s_1 t} + I_2 e^{s_2 t} \tag{1.21}$$

where s_1 and s_2 are the roots of the quadratic equation (1.10). That is,

$$s_{1,2} = \frac{1}{2}\left[-\frac{R^2C + L}{2RCL} \pm \sqrt{\left(\frac{R^2C + L}{2RCL}\right)^2 - \frac{2}{LC}}\right] \tag{1.22}$$

where we use the plus sign for one of the s-values and the minus sign for the other (the choice is arbitrary — whichever way we choose, the values of I_1 and I_2 will adjust to give, in the end, the same final expression for $i(t)$). Depending on the sign of the quantity under the square-root in (1.22), $s_{1,\,2}$ are either real or complex.

Since the complex roots to any algebraic equation with real coefficients always appear as conjugate pairs (this is a very deep theorem in algebra), the two roots to a quadratic are either a complex conjugate pair *or* both roots are real (it is impossible for one root to be real and the other to be complex). It should be obvious *by inspection* of (1.22) that, if both roots are real then both are negative, and that if the roots are a conjugate pair then their real parts are negative. That is, if $s_{1,\,2}$ are a complex conjugate pair then for our circuit we can always write[6]

$$s_{1,2} = \sigma + i\omega,$$

or if $s_{1,\,2}$ are real

$$s_{1,2} = \sigma,$$

where $\sigma < 0$ and $\omega > 0$. Negative real roots are associated with decaying exponential behavior, while complex roots are associated with exponentially damped *oscillatory* (at a frequency determined by the imaginary part of the roots) behavior. (I'll do a detailed example of this second case near the end of this section.) In either case, this means that, even without yet knowing I_1 and I_2, we can conclude

$$\lim_{t\to\infty} i(t) = \frac{1}{R}$$

[6]Notice that i is being used here to denote $\sqrt{-1}$, as well as current. To avoid this double-use, some writers use j to denote $\sqrt{-1}$, but I am going to assume that if you're smart enough to be reading this book, then you're smart enough to know when i is current and when it is $\sqrt{-1}$.

as both exponential terms in $i(t)$ will vanish in the limit (they are *transient*). That is, in the limit of $t \to \infty$ the battery current is a constant (this is called the *steady-state*).

Since we have two constants to determine, we will need to find two equations for I_1 and I_2. One equation is easy to find. Just *before* the switch is closed in Fig. 1.4 the voltage drop across C was given as zero, and the current in L was also given as zero. Since neither of these quantities can change instantly, they must both still be zero just *after* the switch is closed. (If the switch is closed at time $t = 0$, it is standard in electrical engineering to write 'just before' and 'just after' as $t = 0-$ and $t = 0+$, respectively.) Thus, at $t = 0+$ the battery current flows entirely through the two resistors (which are in series), with no voltage drop across the C, and so we have

$$i(0+) = \frac{1}{2R} = \frac{1}{R} + I_1 + I_2$$

or

$$I_1 + I_2 = -\frac{1}{2R}. \tag{1.23}$$

To get our second equation, we need to do a bit more work.

Looking back at (1.3) through (1.6), and setting $u = 1$ for $t > 0$ (which certainly includes $t = 0+$), we have

$$1 = iR + L\frac{di_1}{dt} \tag{1.24}$$

$$L\frac{di_1}{dt} = i_2R + \frac{1}{C}\int_0^t i_2(x)dx \tag{1.25}$$

$$i = i_1 + i_2. \tag{1.26}$$

If we evaluate (1.24) at $t = 0+$, we get

$$1 = i(0+)R + L\frac{di_1}{dt}\Big|_{t=0+} = \frac{1}{2R}R + L\frac{di_1}{dt}\Big|_{t=0+}$$

or,

$$\frac{di_1}{dt}\Big|_{t=0+} = \frac{1}{2L}. \tag{1.27}$$

From (1.26) evaluated at $t = 0+$ we get

$$i(0+) = i_1(0+) + i_2(0+)$$

which, because $i_1(0+) = 0$ (it's the inductor current at $t = 0+$), becomes

$$i_2(0+) = i(0+) = \frac{1}{2R}. \tag{1.28}$$

If we differentiate (1.24) and (1.25) we get

$$0 = R\frac{di}{dt} + L\frac{d^2 i_1}{dt^2}$$

and

$$L\frac{d^2 i_1}{dt^2} = R\frac{di_2}{dt} + \frac{1}{C}i_2$$

which, when combined, gives

$$0 = R\frac{di}{dt} + R\frac{di_2}{dt} + \frac{1}{C}i_2.$$

But since

$$\frac{di}{dt} = \frac{di_1}{dt} + \frac{di_2}{dt}$$

then

$$\frac{di_2}{dt} = \frac{di}{dt} - \frac{di_1}{dt}$$

and so

$$0 = R\frac{di}{dt} + R\frac{di}{dt} - R\frac{di_1}{dt} + \frac{1}{C}i_2$$

or,

$$0 = 2R\frac{di}{dt} - R\frac{di_1}{dt} + \frac{1}{C}i_2.$$

If we now evaluate this last expression at $t = 0+$, and recall our earlier results for $\frac{di_1}{dt}\Big|_{t=0+}$ and $i_2(0+)$ in (1.27) and (1.28), respectively, we have

$$0 = 2R\frac{di}{dt}\Big|_{t=0+} - R\left(\frac{1}{2L}\right) + \frac{1}{C}\left(\frac{1}{2R}\right)$$

or,

$$\frac{di}{dt}\Big|_{t=0+} = \frac{\frac{R}{2L} - \frac{1}{2RC}}{2R} = \frac{1}{4}\left(\frac{1}{L} - \frac{1}{R^2C}\right).$$

But since

$$\frac{di}{dt} = I_1 s_1 e^{s_1 t} + I_2 s_2 e^{s_2 t}$$

we have, for our second equation in I_1 and I_2,

$$I_1 s_1 + I_2 s_2 = \frac{1}{4}\left(\frac{1}{L} - \frac{1}{R^2C}\right). \tag{1.29}$$

With these two equations in two unknowns — (1.23) and (1.29) — it is clear that we can solve for I_1 and I_2 in terms of the circuit component values R, L, and C, and the two values of s in (1.22) which are also completely determined by R, L, and C. To be specific, let's assume some particular values for the components ('sort of' picked at random, but also with the goal of keeping the arithmetic easy for hand calculation, a consideration that will cease to be important once we have access to a computer): $R = 1000$ ohms, $L = 10$ millihenrys, and $C = 0.01$ microfarads. A little 'trick of the transient analyst,' one often useful to keep in mind when working with lots of numbers with exponents, is that if you express resistance in ohms, inductance in microhenrys, and capacitance in microfarads, then time is expressed in microseconds.[7] That is, if we write $R = 10^3$, $L = 10^4$, and $C = 10^{-2}$, then $t = 6$ (for example) means $t = 6$ microseconds. With these numbers for R, L, and C, we have

$$\frac{2}{LC} = \frac{2}{(10^4)(10^{-2})} = 0.02,$$

$$\frac{R^2C + L}{2RCL} = \frac{(10^6)(10^{-2}) + 10^4}{2(10^3)(10^{-2})(10^4)} = 0.1$$

$$s_{1,2} = \frac{1}{2}\left(-0.1 \pm \sqrt{0.01 - 0.02}\right) = \frac{1}{2}\left(-0.1 \pm \sqrt{-0.01}\right) = 0.5(-0.1 \pm i0.1)$$
$$= 0.05(-1 \pm i)$$

and so

$$s_1 = 0.05(-1 + i), s_2 = 0.05(-1 - i).$$

[7]Similarly, using ohms, henrys, and farads gives time in seconds, and using ohms, millihenrys, and millifarads gives time in milliseconds.

Finally,

$$\frac{1}{4}\left(\frac{1}{L}-\frac{1}{R^2C}\right)=\frac{1}{4}\left(\frac{1}{10^4}-\frac{1}{(10^6)(10^{-2})}\right)=0.$$

With these numbers, (1.23) and (1.29) become

$$I_1+I_2=-5\times 10^{-4} \tag{1.30}$$
$$I_1s_1+I_2s_2=0. \tag{1.31}$$

For such a simple system of equations we can solve the first for I_2 in terms of I_1 (or vice versa) and then substitute into the other equation. Much more systematic (especially when there are more than two unknowns), however, is *Cramer's rule*, from the theory of determinants.[8] We'll use that approach later in the book, but for now I'll let you confirm (in any way you wish) that

$$I_1=25\times 10^{-5}(-1+i), I_2=-25\times 10^{-5}(1+i).$$

or, using Euler's identity (see Appendix 1), we finally have

$$i(t)=10^{-3}\left[1-0.5e^{-0.05t}\{\cos(0.05t)+\sin(0.05t)\}\right] \tag{1.32}$$

where the units for *i*(*t*) are amperes and the units for *t* are microseconds.

Figure 1.6 shows a plot of *i*(*t*) for the first 150 μsec after the switch closes in the circuit of Fig. 1.4, and we see that *i*(*t*) starts at 0.5 mA at $t=0+$, rises to slightly more than 1 mA at about $t=60$ μsec, and then declines to a steady-state value of 1 mA.

I'll end this section on an historical note. All that you've read here is standard lore (for electrical engineers, at least) these days, but not so long ago it was cutting-edge science. It was just 165 years ago that the great Irish-born Scottish engineer and mathematical physicist of the nineteenth century British Empire, William Thomson (1824–1907), later Lord Kelvin, gave a talk to the Glasgow Philosophical Society in which the nature of the roots to an equation essentially our (1.22) was of central interest.[9] So struck was Thomson by the dramatic change in the nature of the transient behavior of a circuit, as a function of whether the roots are real or complex, that he offered the following speculation about one of Nature's most spectacular displays of transient electrical fireworks:

[8]Named after the Swiss mathematician Gabriel Cramer (1704–1752), who published it in 1750, but in fact it had appeared in print two years earlier, in the posthumous *Treatise of Algebra* by the Scottish mathematician Colin MacLaurin (1698–1746).

[9]Thomson's remarks in January 1853 to the Society were reprinted a few months later with the title "On Transient Electric Currents" in the June 1853 issue of *The London, Edinburgh and Dublin Philosophical Magazine and Journal of Science* (pp. 393–405). I believe this may have been the very first scientific paper to deal specifically with electrical transients.

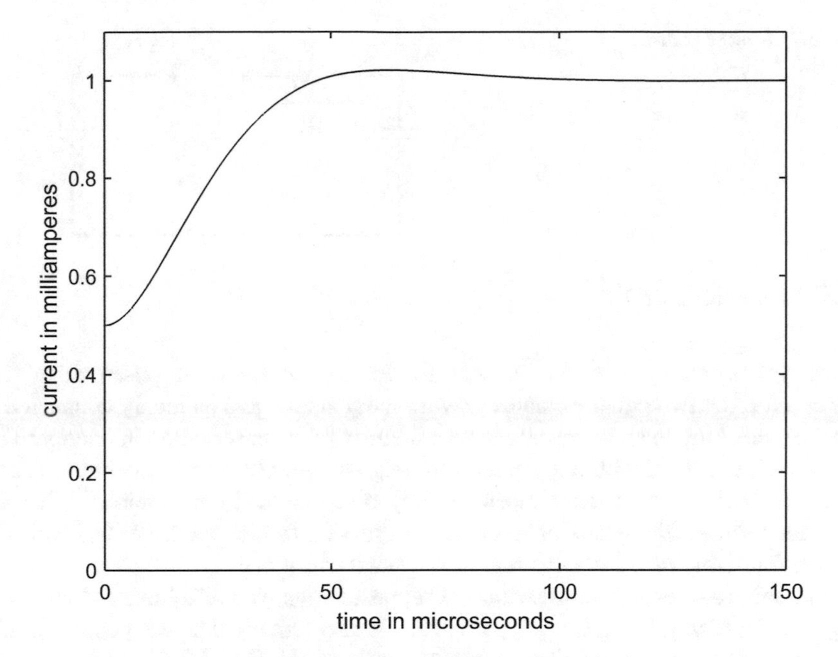

Fig. 1.6 The battery current after switching for the circuit of Fig. 1.4

"It is probable that many remarkable phenomena which have been observed in connection with electrical discharges are due to the oscillatory character which we have thus found to be possessed when the condition [for complex roots] is fulfilled. Thus if the interval of time . . . at which the successive instants when the strength of the current is a maximum follow one another, be sufficiently great, and if the evolution of heat in any part of the circuit by the current during several of its alternations in direction be sufficiently intense to produce visible light, a succession of flashes diminishing in intensity and following one another rapidly at equal intervals will be seen. It appears to me not improbable that double, triple, and quadruple flashes of lightning which I have frequently seen on the continent of Europe, and sometimes, though not so frequently in this country [a reference to Scotland], lasting generally long enough to allow an observer, after his attention is drawn by the first light of the flash, to turn his head round and see distinctly the course of the lightning in the sky, result from the discharge possessing this oscillatory character."

We'll encounter Thomson again, in Chap. 5, when we get to transmission lines in general, and the famous nineteenth century Trans-Atlantic Submarine Telegraph Cable in particular.

Fig. 1.7 A more realistic Fig. 1.5

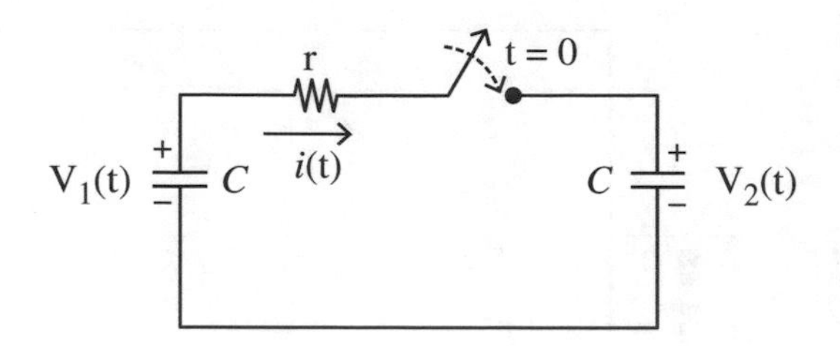

1.5 Puzzle Solution

Have you come-up with an answer to the puzzle question posed Sect. 1.3, concerning the two equal capacitors (with unequal charges) suddenly connected in parallel, resulting in what appears to be a failure of the conservation of energy? This is a classic problem, with one solution of long-standing (the one I'm about to show you), as well as some more recent, purely mathematical explanations[10] that are dependent on subtle details of the impulse function (which we'll get to later). For us, the following *physical* explanation is a satisfactory one.

The failure of energy conservation disappears once we realize that the circuit of Fig. 1.5 is *highly* idealized. In particular, in any circuit that we could actually construct, there would be *some* resistance present. In Fig. 1.7 Fig. 1.5 has been redrawn to show the presence of the resistance r, which we'll take to be arbitrarily small *but not zero*. For $t < 0$ (*before* the switch is closed) $v_1 = V_1$ and $v_2 = V_2$.

Once the switch is closed, the current $i(t)$ flows (let's assume that $V_1 > V_2$, but that assumption is arbitrary and we could just as well assume the opposite). The equation for this circuit is

$$v_1(t) = i(t)r + v_2(t)$$

where, as the voltage drop across the left capacitor is *decreasing*,

$$v_1(t) = V_1 - \frac{1}{C}\int_0^t i(x)dx$$

and, as the voltage drop across the right capacitor is *increasing*,

$$v_2(t) = \frac{1}{C}\int_0^t i(x)dx + V_2.$$

[10] If you are interested in pursuing this (we *won't* in this book), see K. Mita and M. Boufaida, "Ideal Capacitor Circuits and Energy Conservation," *American Journal of Physics*, August 1999, pp. 737–739.

So,

$$V_1 - \frac{1}{C}\int_0^t i(x)dx = ir + \frac{1}{C}\int_0^t i(x)dx + V_2.$$

That is,

$$\frac{2}{C}\int_0^t i(x)dx + ir = V_1 - V_2.$$

Differentiating with respect to time,

$$\frac{2}{C}i + r\frac{di}{dt} = 0$$

or, separating variables,

$$\frac{di}{i} = -\frac{2}{rC}dt.$$

Integrating indefinitely, and writing K as an arbitrary constant,

$$\ln(i) = -\frac{2}{rC}t + K$$

or,

$$i(t) = e^{-\frac{2}{rC}t+K} = e^K e^{-\frac{2}{rC}t} = Ae^{-\frac{2}{rC}t}$$

where $A = e^K$ is a constant. Since

$$i(0) = \frac{V_1 - V_2}{r} = A,$$

then we see that the charge redistribution current is the *exponentially decaying transient*[11]

$$i(t) = \frac{V_1 - V_2}{r}e^{-\frac{2}{rC}t}.$$

[11] This is the generic behavior we'll see over and over whenever we have a capacitor discharging (or charging) through a resistance. Do you see why the rC product has the units of time? That product is called the *time constant* of the transient.

This current will heat r, and that requires energy. How much energy? The answer is

$$\int_0^\infty i^2(t)rdt = \int_0^\infty \frac{(V_1 - V_2)^2}{r^2} e^{-\frac{4}{rC}t} rdt = \frac{(V_1 - V_2)^2}{r} \left\{ -\frac{rC}{4} e^{-\frac{4}{rC}t} \right\} \Big|_0^\infty = C\frac{(V_1 - V_2)^2}{4},$$

which you'll recall is exactly the 'missing' energy. Notice that r, itself, has completely vanished from the analysis, and so our result holds for *any* value of r (*except* for the 'idealized' case of $r = 0$, which is precisely the case that does *not* occur in 'real-life').

Let's turn next to the deeper puzzle of Fig. 3. In the earlier discussion of that circuit in the Preface, I claimed that the current in L_2, for $t < 0$, is undefined, but for discussion here let's specifically say it is zero. (This isn't a crucial point, but it helps keep things simple.) That is, all the battery current that is in L_1 $(= \frac{V}{R})$ flows through the switch that shunts L_1. When the switch is opened at $t = 0$, the current in L_1 begins to decrease and the current in L_2 begins to increase.

The result is to produce a $\frac{di}{dt}$ so large as to create an arc across the switch contacts, an arc which we'll assume extinguishes very quickly by $t = 0+$. Now, at $t = 0-$, that is at just *before* the switch is opened, the total magnetic flux linkage (look back at the end of Sect. 1.1) in both inductors is

$$\phi_L(0-) = \phi_{L_1}(0-) + \phi_{L_2}(0-) = L_1 i(0-) + L_2(0) = L_1 i(0-).$$

What we'll prove next is that the total flux linkage is *conserved* during the switching. That is,

$$\phi_L(0+) = \phi_L(0-).$$

The equation for the circuit of Fig. 3 is

$$V = iR + \frac{d(Li)}{dt},$$

where, you'll note carefully, that the inductor voltage drop is written as $\frac{d(Li)}{dt}$ and *not* as $L\frac{di}{dt}$ because now we are *not* assuming L is a constant. In fact,

$$L = L_1, \ \ t < 0$$
$$L = L_1 + L_2, \ \ t > 0.$$

So,

$$V - iR = \frac{d(Li)}{dt}.$$

Let Δt be the time interval from $t = 0-$ to $t = 0+$ and so, if i_a is in some sense an average value of the current during that tiny interval of time, we can write

$$(V - i_a R)\Delta t = \int_{0-}^{0+} \frac{d(Li)}{dt} dt.$$

Assuming only that, whatever i_a is it is *finite*, then as $\Delta t \to 0$ (that is, the switching is *fast*), the left-hand side goes to zero and so

$$\lim_{\Delta t \to 0} \int_{0-}^{0+} d(Li) = Li(0+) - Li(0-) = 0.$$

That is,

$$\phi_L(0+) - \phi_L(0-) = 0,$$

as claimed.

Using the values for L at $t = 0-$ and at $t = 0+$, we have

$$(L_1 + L_2)i(0+) = L_1 i(0-)$$

and so the current in *both* inductors at $t = 0+$ is

$$i(0+) = \frac{L_1}{L_1 + L_2} i(0-) = \frac{L_1}{L_1 + L_2}\left(\frac{V}{R}\right).$$

Because of the fast switching operation, the currents in both inductors *have* suddenly changed, but this requires what we'll call (later in this book) an *impulsive* voltage drop across the switch which will, as you'll see, be the mathematical explanation for the arc.

1.6 Magnetic Coupling, Part 1

The *coupling* between two inductors L_1 and L_2 (which denote each inductor's so-called *self-inductance*) is represented by M, the *mutual inductance* of the two inductors. $0 \leq M \leq \sqrt{L_1 L_2}$ is a measure of how much the magnetic flux of each inductor links with the coils of the other inductor, there-by inducing an additional voltage drop in the linked inductor. The voltage drop in either one of the coupled inductors therefore depends not only on the current in that inductor, but also on the current in the other inductor. I'll say more on the impact of coupling in the next chapter, and when we get to applying the Laplace transform to transient analysis in Chap. 4, but for now let me show you an interesting, non-intuitive result that we can derive using the trick I showed you in Sect. 1.2, that of *assuming* current/voltage solutions of the form e^{st}.

Imagine an isolated, series R, L, C circuit with initial stored energy (for example, the C is charged). Then, the capacitor is allowed to discharge through the R and L. The equation for the circuit is, with i as the discharge current,

$$iR + L\frac{di}{dt} + \frac{1}{C}\int_0^t i(x)dx = V_0 \tag{1.33}$$

where V_0 is the initial capacitor voltage drop. Then, differentiating,

$$R\frac{di}{dt} + L\frac{d^2i}{dt^2} + \frac{i}{C} = 0.$$

Assuming $i(t) = Ae^{st}$, where A is the amplitude, we have

$$RAse^{st} + LAs^2e^{st} + \frac{Ae^{st}}{C} = 0$$

or, cancelling the Ae^{st} in each term,

$$Rs + Ls^2 + \frac{1}{C} = 0.$$

Solving for s (and remembering my comment in note 6),

$$s = -\frac{R}{2L} \pm i\sqrt{\frac{1}{LC} - \frac{R^2}{4L^2}}.$$

That is, s is complex *if*

$$R < 2\sqrt{\frac{L}{C}}.$$

In this case Euler's identity tells us that the discharge current $i(t)$, for the initially stored energy, will be an exponentially damped (because of the energy loss mechanism of the R) sinusoidal oscillation with the so-called *natural frequency* of[12]

$$f_n = \frac{1}{2\pi}\sqrt{\frac{1}{LC} - \frac{R^2}{4L^2}}\text{hertz}.$$

Now, suppose we have *two R, L, C* series circuits that are magnetically coupled. What influence does the coupling have on the natural frequency of the circuit? This is, in general, not an easy problem to tackle, but it does have a very pretty answer *if* the two circuits have the same natural frequency *when isolated* from each other. When coupled, such a pair of circuits is said to form a *tuned* circuit, and it is easy to

[12]The unit of frequency used to be the natural, obvious *cycles per second* but, deciding that the great German mathematical physicist Heinrich Hertz (1857–1894) had not been suitably honored, the cps was declared by an international committee to be the *hertz* (Hz) in 1960.

show that the coupling results in each of the two series circuits having *two* natural frequencies, one higher and one lower than the single natural frequency f_n each circuit has when isolated. That is, the magnetic coupling causes the single natural frequency of each lone circuit to *split* into *two* possible frequencies.

If we call the currents in the individual (but otherwise identical) R, L, C circuits i_1 and i_2, then, analogous to (1.33), we mathematically write the description of the coupled pair as

$$i_1R + L\frac{di_1}{dt} + \frac{1}{C}\int_0^t i_1(x)dx \pm M\frac{di_2}{dt} = V_0 \tag{1.34}$$

and

$$i_2R + L\frac{di_2}{dt} + \frac{1}{C}\int_0^t i_2(x)dx \pm M\frac{di_1}{dt} = 0, \tag{1.35}$$

where the $\pm$ indicates that the coupling may be of either sign. A positive rate-of-change of current in one circuit could produce either a positive or negative voltage in the other circuit, depending on the relative sense of the windings of each inductor coil. As long as we use the same sign in both equations, the solution will be the same. For the sake of a specific analysis, let's assume the minus sign in both (1.34) and (1.35). Differentiating, we therefore have

$$R\frac{di_1}{dt} + L\frac{d^2i_1}{dt^2} + \frac{i_1}{C} - M\frac{d^2i_2}{dt^2} = 0 \tag{1.36}$$

and

$$R\frac{di_2}{dt} + L\frac{d^2i_2}{dt^2} + \frac{i_2}{C} - M\frac{d^2i_1}{dt^2} = 0. \tag{1.37}$$

Now, assume $i_1 = Ae^{st}$ and $i_2 = Be^{st}$. That is, we'll take the currents in the two circuits as having the same frequency (whatever it is) but not necessarily the same amplitude (A and B are each arbitrary). Then, substituting these assumed currents into (1.36) and (1.37) and making the obvious cancellations,

$$Rs + Ls^2 + \frac{1}{C} - \frac{B}{A}Ms^2 = 0 \tag{1.38}$$

and

$$Rs + Ls^2 + \frac{1}{C} - \frac{A}{B}Ms^2 = 0. \tag{1.39}$$

Thus,

$$\frac{B}{A}Ms^2 = \left(Rs + Ls^2 + \frac{1}{C}\right)$$

and

$$\frac{A}{B}Ms^2 = \left(Rs + Ls^2 + \frac{1}{C}\right).$$

Multiplying these last two results together, we see the arbitrary A and B *vanish*, to give

$$M^2s^4 = \left(Ls^2 + Rs + \frac{1}{C}\right)^2. \tag{1.40}$$

Writing (1.40) as the difference of two squares, we then arrive at

$$\begin{aligned}\left(Ls^2 + Rs + \frac{1}{C}\right)^2 - M^2s^4 = 0 \\ = \left[\left(Ls^2 + Rs + \frac{1}{C}\right) - Ms^2\right]\left[\left(Ls^2 + Rs + \frac{1}{C}\right) + Ms^2\right].\end{aligned}$$

Therefore, either

$$(L - M)s^2 + Rs + \frac{1}{C} = 0$$

or

$$(L + M)s^2 + Rs + \frac{1}{C} = 0.$$

Solving each of these two quadratics for s, we find there are *two* possibilities for the natural frequency of the magnetically coupled pair of circuits:

$$f_{n1} = \frac{1}{2\pi}\sqrt{\frac{1}{(L - M)C} - \frac{R^2}{4(L - M)^2}}$$

and

$$f_{n2} = \frac{1}{2\pi}\sqrt{\frac{1}{(L + M)C} - \frac{R^2}{4(L + M)^2}}.$$

One of these two frequencies will always be greater than f_n, while the other will be smaller. Which is which, however, depends on the values of the circuit components (see Problem 1.8). When transients exist in magnetically coupled circuits, *both* frequencies will generally be present, and you'll see examples of that situation in Chaps. 2 and 4. Notice that as $M \to 0$ (the two circuits become *uncoupled*) both f_{n1} and f_{n2} approach f_n.

Problems

1.1 In Fig. 1.8 the switch has been closed for a long time. At time $t = 0$ it is opened.

(a) Just before that (at time $t = 0-$), what are the currents in the two equal-valued inductors, and what is the voltage drop across the capacitor?
(b) For $t > 0$, determine under what condition the circuit will oscillate.

Fig. 1.8 Problem 1.1

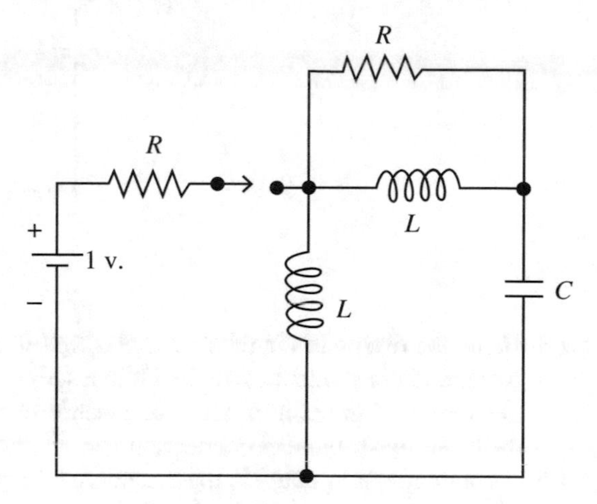

1.2 In the circuit shown in Fig. 1.9, the capacitor is charged to V_0 volts. At time $t = 0$ the switch is closed and the capacitor begins to discharge. Show that the circuit will oscillate only if the value of R is at least some minimum value which is a function of the values of L and C. Hint: with $v(t)$ as the voltage drop across the parallel R and C, show that

$$LC\frac{d^2v}{dt^2} + \left(\frac{L}{R}\right)\frac{dv}{dt} + v = 0$$

and assume v is of the form e^{st}. Under what conditions does s have an imaginary part?

Fig. 1.9 Problem 1.2

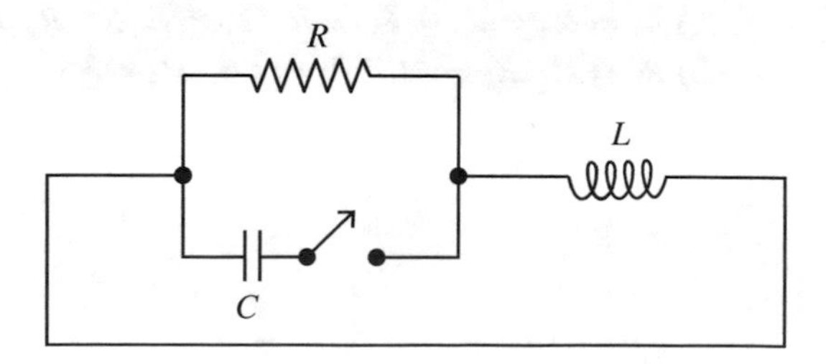

1.3 In the circuit shown in Fig. 1.10 the switch is closed at $t = 0$. If the battery voltage is 1 volt, what is the voltage drop across each capacitor after a long time has passed?

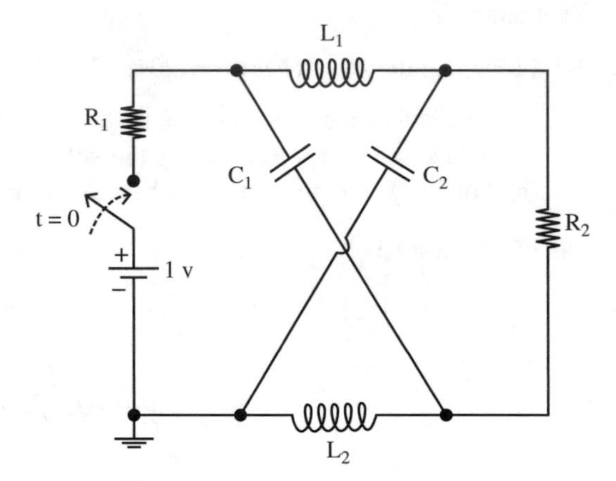

Fig. 1.10 Problem 1.3

1.4 Using the discussion in the text on the conservation of magnetic flux linkage for two inductors suddenly switched into a series connection as a guide, construct a derivation of the conservation of electric charge for two capacitors suddenly switched into a parallel connection.

1.5 As pointed out in note 10, the discharge of a capacitor through a resistance will be, in general, an exponential decay of the form $e^{-t/\tau}$, where τ is the product of the resistance and the capacitance. Since the exponent must be dimensionless (see the next problem) it follows that that product must have the dimensions of time. Show that this is so, using the fundamental relations of (a) current has the units of charge/time, (b) resistance has the units of voltage/current, and (c) voltage has the units of charge/capacitance.

1.6 Why must the power of an exponential be dimensionless? Hint: suppose the exponent did have dimensions, and then consider the power series expansion of the exponential. See any difficulties?

1.7 In Fig. 1.11 the switch has been closed for a long time, and the circuit is in its steady-state. Then, at $t = 0$, the switch is opened. What is the voltage drop across the switch contacts at $t = 0-$ and at $t = 0+$? Consider three cases: (1) $R_1 = R_2 = R_3 = R_4 = R$, (2) $R_1 = R,\ R_2 = 2R,\ R_3 = 3R,\ R_4 = 4R$, and (3) $R_1 = R,\ R_2 = \frac{1}{2}R,\ R_3 = \frac{1}{3}R,\ R_4 = \frac{1}{4}R$.

Fig. 1.11 Problem 1.7

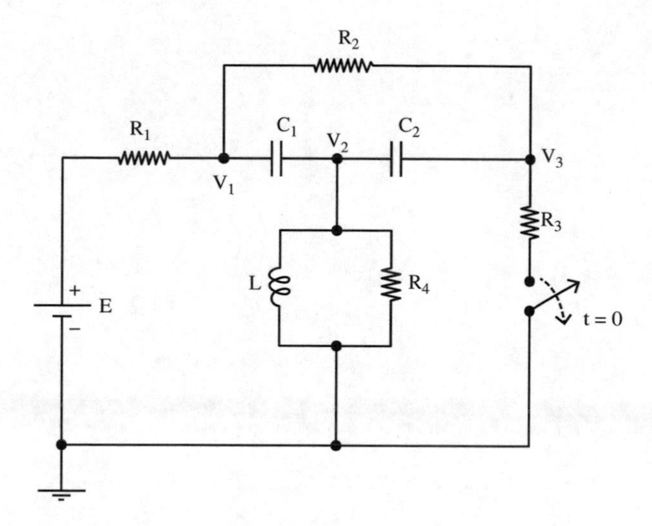

1.8 For a magnetically coupled tuned circuit, it is obvious that for the case of $R = 0$ we have

$$f_{n1} > f_n > f_{n2}.$$

What is the required condition for this to still hold in the case of $= \sqrt{\frac{L}{C}}$?

Chapter 2
Transients in the Time Domain

2.1 Sometimes You Don't Need a Lot of Math

The circuit in Fig. 2.1 has had the switch closed for a long time, and then it is opened (as shown) at time $t = 0$. What is the capacitor voltage v_c (at point **a**) for time $t \geq 0$? We could start to answer this question by writing down Kirchhoff's equations and then doing some (maybe more than a little) algebra, but instead let's see if we can use the ideas from Chap. 1 to arrive at the solution *without* doing a lot of algebra.

To start, pay particular attention to how the battery is positioned. Its positive terminal is connected directly to ground which, *by definition*, is always at zero potential. Thus, the negative terminal (point **c**) is at potential –*E*. That is, point **c** is at *E* volts *below ground*. So, just before the switch is opened the capacitor is charged to $\frac{E}{2}$ volts, with point **a** *below* ground, i.e., $v_c(0-) = -\frac{E}{2}$. That's because the middle R_1 is shorted by the closed switch and, as there is no current in R_2 and the *C* when the switch has been closed for a long time, then point **b** is the mid-point of a voltage divider formed by the other two R_1 resistors which puts point **b** midway between –*E* volts and ground.

Now, since a capacitor voltage drop can't change instantly, we have $v_c(0+) = -\frac{E}{2}$. Then, with the switch open, a long time later all will be as before except that point **b** is now part of a voltage divider made from *three* R_1 resistors and so point **b**'s potential will be at $\frac{E}{3}$ volts below ground when $t = \infty$. That is, $v_c(\infty) = -\frac{E}{3}$. So, *C* discharges from $= -\frac{E}{2}$ to $-\frac{E}{3}$, and the discharge will be exponential (look back at note 10 in Chap. 1) with some time constant $\tau = RC$ where *R* is a resistance determined by the values of R_1 and R_2 (will figure-out what *R* is in just a moment). That is,

$$v_c(t) = -\frac{E}{3}\left(1 + \frac{1}{2}e^{-\frac{t}{\tau}}\right).$$

P. J. Nahin, *Transients for Electrical Engineers*,
https://doi.org/10.1007/978-3-319-77598-2_2

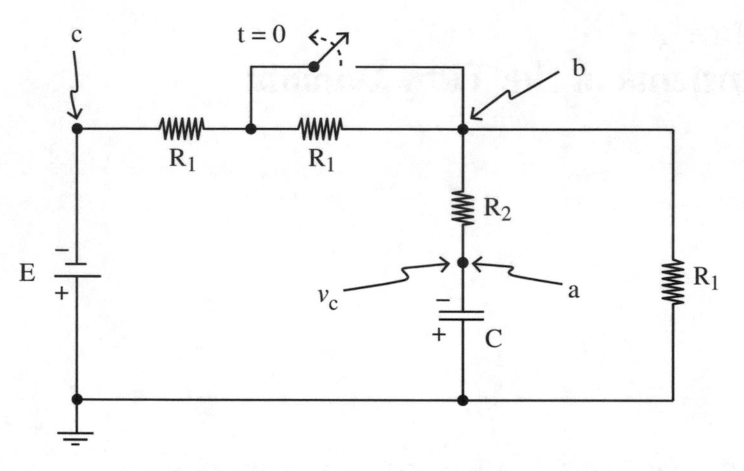

Fig. 2.1 What is $v_c(t)$ at point **a** for $t \geq 0$?

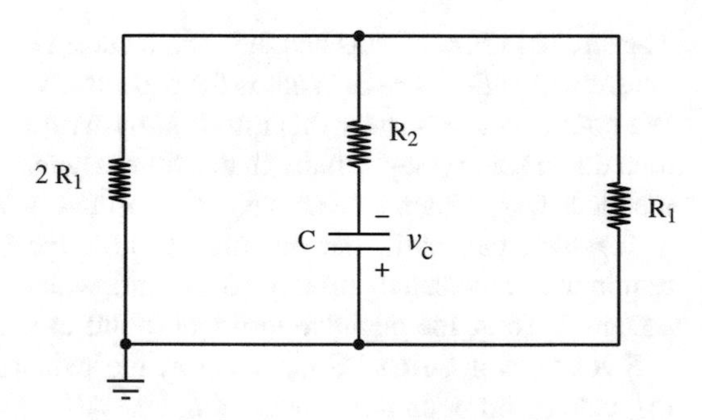

Fig. 2.2 The discharge circuit for $t \geq 0$

Okay, what's τ (that is, what's R)? Once the switch is closed, the circuit 'looks' as shown in Fig. 2.2 where the *ideal* battery has been replaced with its internal resistance (zero).

That is, C discharges through R_2 in series with the combination of $2R_1$ in parallel with R_1. Thus,

$$R = R_2 + \frac{2R_1^2}{2R_1 + R_1} = R_2 + \frac{2R_1}{3} = \frac{3R_2 + 2R_1}{3}$$

and so

$$\tau = \frac{3R_2 + 2R_1}{3} C$$

and thus, just like that, we have

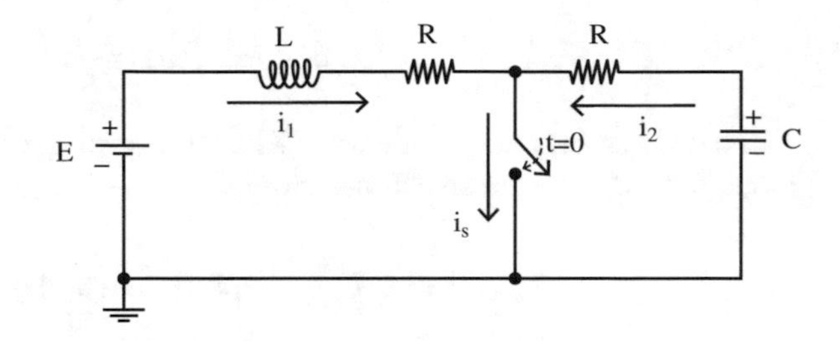

Fig. 2.3 What is the switch current for $t \geq 0$?

$$v_c(t) = -\frac{E}{3}\left(1 + \frac{1}{2}e^{-\frac{3t}{(3R_2+2R_1)C}}\right), t \geq 0. \tag{2.1}$$

Is the physical reasoning I've just taken you through to arrive at (2.1) correct? Perhaps (you might wonder) there could be some subtle error somewhere? Well, the result *is* correct, as a formal mathematical analysis would show. It's easier to do that formal analysis in the transform domain than it is to do in the time domain, and so you might want to come back to this problem again, later, when we get to the Laplace transform.

2.2 An Interesting Switch-Current Calculation

In the circuit of Fig. 2.3 the switch has been open (as shown) for a long time. At time $t = 0$ the switch is closed. What is $i_s(t)$, the switch current, for $t \geq 0$? Notice that there is no need to be concerned about a switching arc in this problem, as the switch is not interrupting an inductor current. Now, before we write any mathematics at all, we can immediately make the following observation: $i_s(0+) = i_s(\infty)$.

To see that this is so, notice that at $t = 0-$ (just before we close the switch) C is charged to E volts[1] and so there is zero current in the inductor. This last point means that at time $t = 0+$ (just after we close the switch) the inductor current will still be zero. So, the only source of current in the switch at the instant the switch closes is from the discharge of C. Since the capacitor voltage drop at $t = 0-$ is E, then at $t = 0$ + the capacitor voltage drop is still E and, since the capacitor discharge current at $t = 0+$ is $\frac{E}{R}$ then $i_s(0+)$ is $\frac{E}{R}$. Next, after a long time has passed the only current flowing is that of the battery, i_1, through the series L and R and the switch (the closed switch, of course, completely short-circuits the series R and C). The battery current is then clearly $\frac{E}{R}$, and so $i_s(\infty) = \frac{E}{R}$. Thus, as claimed,

[1]Since the bottom terminal of C is connected to ground, the potential of the top terminal of C is the voltage drop across C.

$$i_s(0+) = i_s(\infty) = \frac{E}{R}. \tag{2.2}$$

To find $i_s(t)$ in general, let's write Kirchhoff's voltage laws around each of the *two* loops formed once the switch has closed:

$$-E + L\frac{di_1}{dt} + i_1 R = 0, i_1(0+) = 0 \tag{2.3}$$

and

$$-E + \frac{1}{C}\int_0^t i_2(u)du + i_2 R = 0, i_2(0+) = \frac{E}{R} \tag{2.4}$$

where the $-E$ in (2.3) is the voltage *rise* (going clockwise) through the battery, and the $-E$ in (2.4) is the voltage *rise* (going counter-clockwise) due to the initial charge on the capacitor. Thus, differentiating (2.4) with respect to t,

$$\frac{1}{C}i_2(t) + R\frac{di_2}{dt} = 0$$

or,

$$\frac{di_2}{dt} + \frac{1}{RC}i_2(t) = 0. \tag{2.5}$$

Mathematicians call (2.5) a *separable* equation because we can separate the variables i_2 and t to opposite sides of the equality to write

$$\frac{di_2}{i_2} = -\frac{1}{RC}dt$$

which immediately integrates to

$$\ln(i_2) = -\frac{t}{RC} + k$$

where k is some constant. Since

$$i_2(0+) = \frac{E}{R}$$

we have

$$\ln\left(\frac{E}{R}\right) = k.$$

Thus,

$$\ln(i_2) = -\frac{t}{RC} + \ln\left(\frac{E}{R}\right)$$

or

$$\ln(i_2) - \ln\left(\frac{E}{R}\right) = -\frac{t}{RC} = \ln\left(\frac{i_2}{E/R}\right)$$

or

$$i_2 = \frac{E}{R}e^{-t/RC} \tag{2.6}$$

which you may have already anticipated, using the same argument we used in the previous section. Also, from (2.3) we have

$$\frac{di_1}{dt} + \frac{R}{L}i_1 = \frac{E}{L}. \tag{2.7}$$

To solve (2.7), we use the idea of combining its constant solution (*constant* means $\frac{di_1}{dt} = 0$) to the *time varying* solution of the homogeneous version of (2.7):

$$\frac{di_1}{dt} + \frac{R}{L}i_1 = 0. \tag{2.8}$$

That constant solution of (2.7) is seen, by inspection, to be $i_1 = \frac{E}{R}$. The time-varying solution to (2.8) follows immediately from our earlier solution for the same equation for i_2 (where now $\frac{R}{L}$ plays the role of $\frac{1}{RC}$). That is, the time-varying component of i_2 is $\frac{E}{R}e^{-\frac{R}{L}t}$, and so the complete $i_2(t)$ is

$$i_2(t) - \frac{E}{R} - \frac{E}{R}e^{-\frac{R}{L}t} \tag{2.9}$$

where we *subtract* the two individual solutions because $i_1(0+) = 0$. Since the switch current is, from Kirchhoff's node law for currents, $i_s(t) = i_1(t) + i_2(t)$, we have

$$i_s(t) = \frac{E}{R}e^{-t/RC} + \frac{E}{R} - \frac{E}{R}e^{-\frac{R}{L}t}$$

or,

$$i_s(t) = \frac{E}{R}\left[1 + e^{-t/RC} - e^{-\frac{R}{L}t}\right], t \geq 0. \tag{2.10}$$

Notice that (2.10) says that $i_s(0+) = i_s(\infty) = \frac{E}{R}$, just we argued *physically* at the start. The mathematical treatment does tell us something else quite interesting, however, that we didn't know before:

$$i_s(t) = \frac{E}{R}, \text{for } all\ t,$$

if $\frac{R}{L} = \frac{1}{RC}$. That is, under the particular condition of $R = \sqrt{\frac{L}{C}}$ the switch current will be *unvarying*, even though elsewhere in the circuit the i_1 and i_2 currents are individually continually changing.

You can tell from (2.10) that, depending on which exponential term goes to zero faster as $t \to \infty$, $i_s(t)$ will have either a maximum or a minimum value. For example, if $\frac{1}{RC} > \frac{R}{L}$ then the $e^{-t/RC}$ term will be dominated by the $-e^{-\frac{R}{L}t}$ term, and so $i_s(t)$ will *decrease* from its initial $\frac{E}{R}$ value before it ends-up back at its final $\frac{E}{R}$ value. That is, if $\frac{1}{RC} > \frac{R}{L}$ then $i_s(t)$ should have a *minimum* value at some time $t = T_{min}$. Similarly, if $\frac{1}{RC} < \frac{R}{L}$ then $i_s(t)$ should have a *maximum* value at some time $t = T_{max}$.

To find the extrema of $i_s(t)$ set $\frac{di_s}{dt} = 0$, getting

$$\frac{di_s}{dt} = \frac{E}{R}\left[\frac{R}{L}e^{-\frac{R}{L}t} - \frac{1}{RC}e^{-t/RC}\right]$$

and this is zero if

$$\frac{R}{Le^{\frac{R}{L}t}} = \frac{1}{RCe^{t/RC}}$$

or

$$\frac{e^{t/RC}}{e^{\frac{R}{L}t}} = \frac{L}{R^2C} = e^{\frac{1}{RC} - \frac{R}{L}t}.$$

This is quickly solved for t to give

$$t = \frac{\ln\left(\frac{1/RC}{R/L}\right)}{\frac{1}{RC} - \frac{R}{L}}.$$

Notice that if $\frac{1}{RC} > \frac{R}{L}$ then both numerator and denominator are positive and the time T_{min} is

$$T_{min} = \frac{\ln\left(\frac{1/RC}{R/L}\right)}{\frac{1}{RC} - \frac{R}{L}}$$

Just to see how the numbers go, suppose $E = 100$ volts, $L = 100$ mH, and $C = 0.01$ microfarads. Then

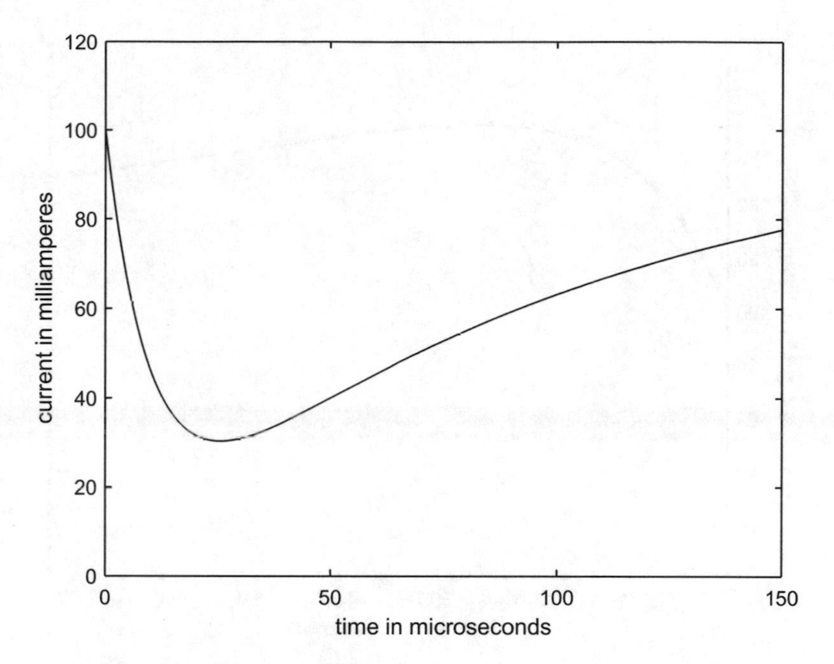

Fig. 2.4 When the switch current has a minimum

$$\frac{1}{RC} = \frac{1}{(10^3)(10^{-8})} = \frac{1}{10^{-5}} = 10^5$$

and

$$\frac{R}{L} = \frac{10^3}{10^{-1}} = 10^4$$

and so $\frac{1}{RC} > \frac{R}{L}$ and $i_s(t)$ will have a minimum at

$$T_{min} = \frac{\ln\left(\frac{10^5}{10^4}\right)}{10^5 - 10^4} = \frac{\ln(10)}{90,000} = 25.58\ \mu sec.$$

The switch current at $t = T_{min}$ is

$$i_s(T_{min}) = \frac{100}{1,000}\left[1 + e^{-25.58x10^{-6}x10^5} - e^{-25.58x10^{-6}x10^4}\right]$$
$$= 100mA\left[1 + e^{-2.558} - e^{-0.2558}\right] = 100ma[1 + 0.077 - 0.774] = 30.3\ mA.$$

Figure 2.4 shows the switch current, given by (2.10), for the first 150 microseconds after the switch is closed, and we see that the minimum is a rather deep one, with a relatively slow recovery compared to how quickly the minimum is achieved.

Suppose that C is changed to 1 microfarad. Then

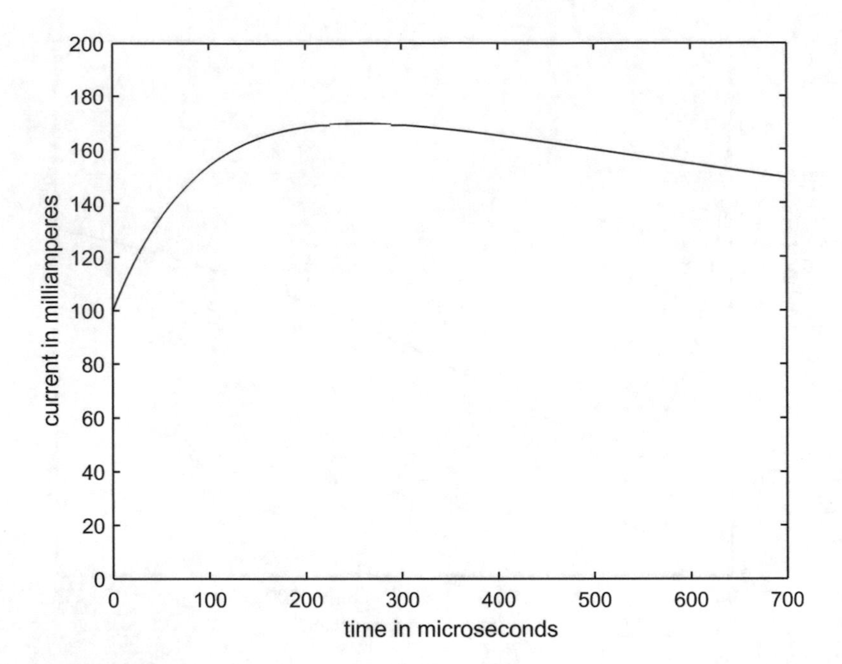

Fig. 2.5 When the switch current has a maximum

$$\frac{1}{RC} = \frac{1}{(10^3)(10^{-6})} = \frac{1}{10^{-3}} = 10^3$$

and

$$\frac{1}{RC} < \frac{R}{L}$$

and so the switch current will have a maximum at

$$T_{max} = \frac{\ln\left(\frac{10^3}{10^4}\right)}{10^3 - 10^4} = \frac{\ln(0.1)}{-9,000} = \frac{\ln(10)}{9,000} = 255.8\ \mu sec.$$

The switch current at $t = T_{max}$ is

$$i_s(T_{max}) = 100mA\left[1 + e^{-255.8x10^{-6}x10^3} - e^{-255.8x10^{-6}x10^4}\right]$$
$$= 100mA\left[1 + e^{-.2558} - e^{-2.558}\right] = 100ma[1 + 0.774 - 0.077] = 169.7\ mA.$$

Figure 2.5 shows the switch current, given by (2.10), for the first 700 microseconds after the switch is closed. The maximum is a very broad one.

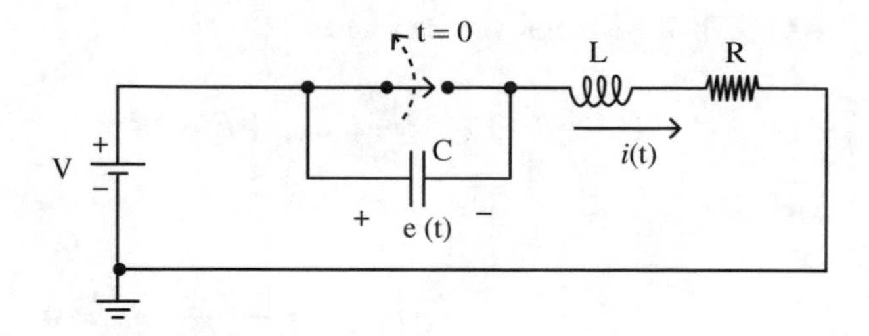

Fig. 2.6 Will an arc form when the switch opens?

2.3 Suppressing a Switching Arc

For our next time domain example, let's return to the issue that inspired this book: the arc that tends to form when an inductive circuit is mechanically interrupted, such as by a circuit-breaker. One common way to attempt to suppress such an arc is to place a capacitor C across the switch contacts, as shown in Fig. 2.6.

It's easy to see, physically, what C does. With the switch closed, C is 'shorted' by the switch and so will have zero charge when the switch is opened at $t = 0$; C then presents a path for the inductor's current, and so, with zero voltage drop across the switch at $t = 0+$ (the voltage drop across the uncharged capacitor), there is (at least initially) no arc. The inductor current will, of course, cause the capacitor voltage drop, $e(t)$, to move away from zero, perhaps in an oscillatory way (depending on the particular values of R, L, and C), and so the question of interest for us is: just what *is* $e(t)$ for $t > 0$?

It's physically clear that, after a long time, the final capacitor voltage drop will be V. That is,

$$\lim_{t\to\infty} e(t) = V. \tag{2.11}$$

When the circuit has been in operation *as shown* in the figure for a long time, *before* we open the switch, the steady-state current is the constant $(t) = \frac{V}{R}$, as the inductor would have no voltage drop. The capacitor has zero voltage drop, too, of course, as it is by-passed by the closed switch. Then, at time $t = 0$, the switch is opened. Without the C there would indeed be an arc because the inductor 'resists' a change in the current. But with the C in the circuit by-passing the open switch, matters are now quite different. The C gives the inductor current an alternative path to that of 'jumping the switch,' a path that keeps the initial voltage drop across the switch at zero.

The Kirchhoff voltage loop equation of the circuit is

$$\frac{1}{C}\int_0^t i(x)dx + L\frac{di}{dt} + iR = V, i(0+) = \frac{V}{R}, \tag{2.12}$$

At $t = 0+$ the integral vanishes and

$$L\frac{di}{dt}\Big|_{t=0+} + i(0+)R = V$$

and so

$$\frac{di}{dt}\Big|_{t=0+} = \frac{V - i(0+)R}{L} = 0. \tag{2.13}$$

A second differentiation of the Kirchhoff equation gives

$$\frac{1}{C}i + L\frac{d^2i}{dt^2} + R\frac{di}{dt} = 0, i(0+) = \frac{V}{R}, \frac{di}{dt}\Big|_{t=0+} = 0.$$

Next, assume $i(t) = Ie^{st}$. Then

$$\frac{I}{C}e^{st} + LIs^2e^{st} + RsIe^{st} = 0$$

or,

$$\frac{1}{C} + Ls^2 + Rs = 0$$

or,

$$s^2 + \frac{R}{L}s + \frac{1}{LC} = 0$$

and so

$$s = \frac{-\frac{R}{L} \pm \sqrt{\left(\frac{R}{L}\right)^2 - \frac{4}{LC}}}{2} = -\frac{R}{2L} \pm \sqrt{\left(\frac{R}{2L}\right)^2 - \frac{1}{LC}}.$$

Thus,

$$i(t) = I_1e^{s_1t} + I_2e^{s_2t} \tag{2.14}$$

where

$$s_1 = -\frac{R}{2L} - \sqrt{\left(\frac{R}{2L}\right)^2 - \frac{1}{LC}}, s_2 = -\frac{R}{2L} + \sqrt{\left(\frac{R}{2L}\right)^2 - \frac{1}{LC}}.$$

Using $i(0+)$ from (2.12), we have

$$I_1 + I_2 = \frac{V}{R} . \tag{2.15}$$

And since

$$\frac{di}{dt} = I_1 s_1 e^{s_1 t} + I_2 s_2 e^{s_2 t},$$

then from (2.13) we have

$$I_1 s_1 + I_2 s_2 = 0. \tag{2.16}$$

Using Cramer's rule on the simultaneous equations of (2.15) and (2.16), the system determinant is

$$D = \begin{vmatrix} 1 & 1 \\ s_1 & s_2 \end{vmatrix} = s_2 - s_1$$

and so we can therefore immediately write

$$I_1 = \frac{\begin{vmatrix} \frac{V}{R} & 1 \\ 0 & s_2 \end{vmatrix}}{D} = \frac{\frac{V}{R} s_2}{D} \tag{2.17}$$

and

$$I_2 = \frac{\begin{vmatrix} 1 & \frac{V}{R} \\ s_1 & 0 \end{vmatrix}}{D} = -\frac{\frac{V}{R} s_1}{D}. \tag{2.18}$$

So,

$$i(t) = \frac{\frac{V}{R} s_2}{D} e^{s_1 t} - \frac{\frac{V}{R} s_1}{D} e^{s_2 t}$$

or,

$$i(t) = \frac{V}{RD}\left[s_2 e^{s_1 t} - s_1 e^{s_2 t} \right]. \tag{2.19}$$

The voltage drop across the open switch is

$$e(t) = \frac{1}{C}\int_0^t i(x)dx = \frac{V}{RDC}\left[s_2 \int_0^t e^{s_1 x}dx - s_1 \int_0^t e^{s_2 x}dx \right]$$

which, after doing the easy integrals, gives

$$e(t) = \frac{V}{RDC}\left[\frac{s_2}{s_1} e^{s_1 t} - \frac{s_2}{s_1} - \frac{s_1}{s_2} e^{s_2 t} + \frac{s_1}{s_2} \right]. \tag{2.20}$$

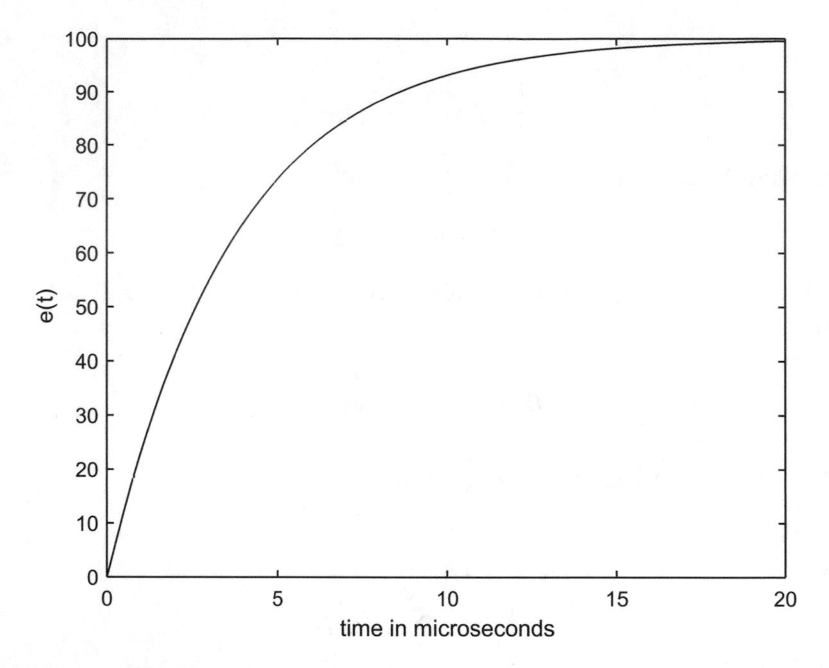

Fig. 2.7 The switch voltage for real values of *s*

To see what (2.20) 'looks like' when plotted, there are two distinct cases to consider: when s_1 and s_2 are real, and when they are complex. I'll discuss each case in turn.

When $\left(\frac{R}{2L}\right)^2 > \frac{1}{LC}$ then both s_1 and s_2 are real. And, it's important to note, negative, which means the transient exponential terms decay with time, rather than grow (which is physically impossible). For example, suppose $R = 400$ ohms, $L = 100\ \mu$H, and $C = 0.01\ \mu$F. Then,

$$\frac{1}{LC} = \frac{1}{\left(10^{-4}\right)\left(10^{-8}\right)} = 10^{12}$$

while

$$\left(\frac{R}{2L}\right)^2 = \left(\frac{400}{2 \times 10^{-4}}\right)^2 = 4 \times 10^{12}.$$

Figure 2.7 shows $e(t)$ for the first 20 microseconds after the switch is closed, for a battery voltage of $V = 100$ volts.

Next, keep L and C the same but reduce R to 100 ohms. Then

$$\left(\frac{R}{2L}\right)^2 = \left(\frac{100}{2\times 10^{-4}}\right)^2 = 0.25\times 10^{12}$$

and now we'll have complex values for s_1 and s_2. In particular,

$$s_1 = -\frac{R}{2L} - i\sqrt{\frac{1}{LC} - \left(\frac{R}{2L}\right)^2} = -\alpha - i\omega$$

and

$$s_2 = -\frac{R}{2L} + i\sqrt{\frac{1}{LC} - \left(\frac{R}{2L}\right)^2} = -\alpha + i\omega$$

where

$$\alpha = \frac{R}{2L}, \omega = \sqrt{\frac{1}{LC} - \left(\frac{R}{2L}\right)^2}.$$

Also,

$$D = s_2 - s_1 = i2\omega.$$

Thus, (2.20) becomes

$$e(t) = \frac{100}{RCi2\omega}\left[\frac{-\alpha+i\omega}{-\alpha-i\omega}e^{(-\alpha-i\omega)t} - \frac{-\alpha+i\omega}{-\alpha-i\omega} - \frac{-\alpha-i\omega}{-\alpha+i\omega}e^{(-\alpha+i\omega)t} + \frac{-\alpha-i\omega}{-\alpha+i\omega}\right]$$

which, with a bit of algebra and Euler's identity, reduces to

$$e(t) = \frac{100}{RC\omega(\alpha^2+\omega^2)}\left[2\alpha\omega - e^{-\alpha t}\left\{\left(\alpha^2-\omega^2\right)\sin(\omega t) + 2\alpha\omega\cos(\omega t)\right\}\right]. \quad (2.21)$$

As partial checks on (2.21), notice that $e(0+) = 0$, and that $e(\infty) = 100$, which are correct.

Figure 2.8 shows $e(t)$ for the first 20 microseconds after the switch is closed, and it looks *much* different from before (in Fig. 2.7). Besides being oscillatory, notice that the peak value of $e(t)$ now exceeds the battery voltage of 100 volts.

2.4 Magnetic Coupling, Part 2

In Chap. 1 we examined a special case of magnetically coupled circuits: that of the two circuits being *tuned*, with each having the same *isolated* natural frequency. In this section we'll treat a different special case. Now each of the two circuits can have

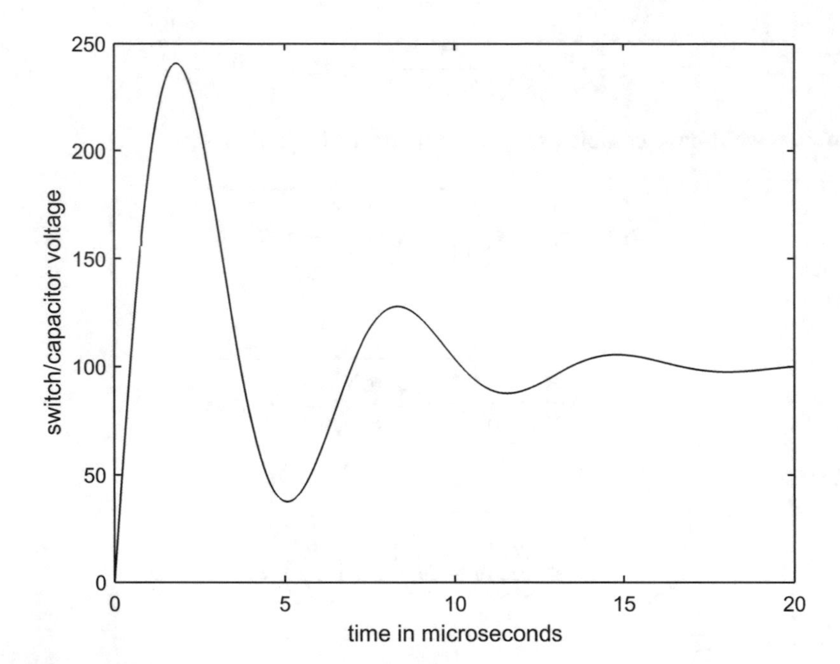

Fig. 2.8 The switch voltage for complex values of *s*

Fig. 2.9 Magnetic coupling with zero energy loss

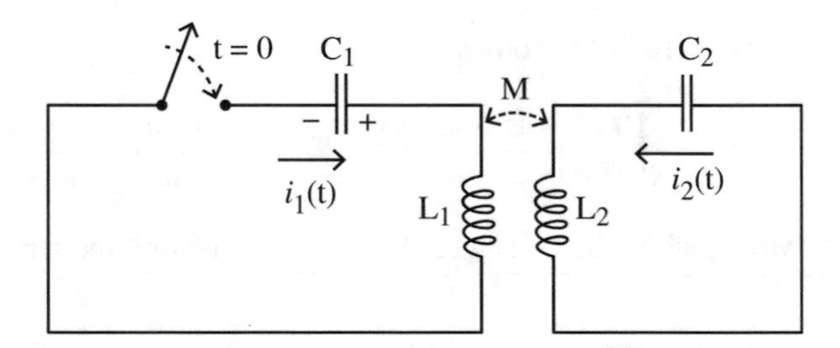

any natural frequency desired (*not* necessarily the same for each circuit) but, to keep the analysis mathematically under control, we'll assume (unlike in Chap. 1) the resistance in each circuit is zero, as shown in Fig. 2.9. This is not physically possible in real circuits, of course, but for coupled circuits that have 'low' resistances it may be a reasonable first approximation, at least for an initial period.

When we get to Chap. 4, after having developed the Laplace transform in Chap. 3, we'll be able to handle the general case where energy loss *is* present in both circuits, but for now our problem here is to calculate $i_1(t)$ and $i_2(t)$ — the so-called *primary* and *secondary* currents, respectively — for $t \geq 0$ with no energy loss. As in Sect. 1.6, C_1 has an initial charge (stored energy).

Is it 'obvious' to you what $i_1(t)$ and $i_2(t)$ are? It would seem that both would oscillate, and that those oscillations should not decay with time since there is no energy dissipation mechanism present. What I think *will* surprise you, however, is just how complicated (because of the magnetic coupling) the calculation of those oscillations can be. The reason for this complication is actually pretty easy to understand: when the charged capacitor C_1 begins to discharge and create $i_1(t)$, that current induces a voltage in L_2 which starts $i_2(t)$ which then induces a voltage back in L_1 which of course influences $i_1(t)$; that influence is then coupled back into L_2 and ... on and on it goes, with energy exchanged back-and-forth between the two circuits in a mutually interacting manner that almost instantly becomes impossible for a human mind to intuitively follow. But our *mathematics* isn't so easily overwhelmed, and it *can* follow — if we are careful — the evolution of the two currents. This calculation will, however, be an illustration of just about the limit of our willingness to do a pure time domain analysis.

Looking back at (1.38) and (1.39), if we set $R = 0$, and write L_1 and L_2, and C_1 and C_2, for the inductors and capacitors in the two circuits, we have

$$L_1 s^2 + \frac{1}{C_1} - \frac{B}{A} M s^2 = 0$$

and

$$L_2 s^2 + \frac{1}{C_2} - \frac{A}{B} M s^2 = 0$$

or,

$$\frac{B}{A} M s^2 = L_1 s^2 + \frac{1}{C_1}$$

and

$$\frac{A}{B} M s^2 = L_2 s^2 + \frac{1}{C_2}.$$

Thus,

$$M^2 s^4 = \left(L_1 s^2 + \frac{1}{C_1}\right)\left(L_2 s^2 + \frac{1}{C_2}\right) = L_1 L_2 s^4 + \left(\frac{L_1}{C_2} + \frac{L_2}{C_1}\right) s^2 + \frac{1}{C_1 C_2}$$

or,

$$(L_1 L_2 - M^2) s^4 + \left(\frac{L_1}{C_2} + \frac{L_2}{C_1}\right) s^2 + \frac{1}{C_1 C_2} = 0.$$

This is a quadratic in s^2, and so we can write by inspection that

$$s = \pm \sqrt{\frac{-\left(\frac{L_1}{C_2} + \frac{L_2}{C_1}\right) \pm \sqrt{\left(\frac{L_1}{C_2} + \frac{L_2}{C_1}\right)^2 - 4\,\frac{L_1 L_2 - M^2}{C_1 C_2}}}{2(L_1 L_2 - M^2)}}.$$

There are four values of s, corresponding to the four ways of specifying the plus/minus signs. Notice, carefully, that all four values are pure imaginary, as the quantity under the inner square-root is positive because, first,

$$\left(\frac{L_1}{C_2}+\frac{L_2}{C_1}\right)^2-4\frac{L_1L_2-M^2}{C_1C_2}=\frac{{L_1}^2}{{C_2}^2}+2\frac{L_1L_2}{C_1C_2}+\frac{{L_2}^2}{{C_1}^2}-4\frac{L_1L_2}{C_1C_2}+4\frac{M^2}{C_1C_2}$$
$$=\frac{{L_1}^2}{{C_2}^2}-2\frac{L_1L_2}{C_1C_2}+\frac{{L_2}^2}{{C_1}^2}+4\frac{M^2}{C_1C_2}=\left(\frac{L_1}{C_2}-\frac{L_2}{C_1}\right)^2+4\frac{M^2}{C_1C_2}>0.$$

In addition, the quantity under the outer square-root is negative because

$$\left(\frac{L_1}{C_2}+\frac{L_2}{C_1}\right)>\sqrt{\left(\frac{L_1}{C_2}+\frac{L_2}{C_1}\right)^2-4\frac{L_1L_2-M^2}{C_1C_2}}$$

since $0\leq M\leq\sqrt{L_1L_2}$. These four values of s are, of course, in the form of two conjugate pairs, and so we have

$$s_1=i\omega_1, s_2=-i\omega_1, s_3=i\omega_2, s_4=-i\omega_2$$

where

$$\omega_1=\sqrt{\frac{\left(\frac{L_1}{C_2}+\frac{L_2}{C_1}\right)+\sqrt{\left(\frac{L_1}{C_2}+\frac{L_2}{C_1}\right)^2-4\frac{L_1L_2-M^2}{C_1C_2}}}{2\left(L_1L_2-M^2\right)}} \tag{2.22}$$

and

$$\omega_2=\sqrt{\frac{\left(\frac{L_1}{C_2}+\frac{L_2}{C_1}\right)-\sqrt{\left(\frac{L_1}{C_2}+\frac{L_2}{C_1}\right)^2-4\frac{L_1L_2-M^2}{C_1C_2}}}{2\left(L_1L_2-M^2\right)}}. \tag{2.23}$$

These results tell us that

$$i_1(t)=K_1e^{i\omega_1t}+K_2e^{-i\omega_1t}+K_3e^{i\omega_2t}+K_4e^{-i\omega_2t}$$

and

$$i_2(t)=K_5e^{i\omega_1t}+K_6e^{-i\omega_1t}+K_7e^{i\omega_2t}+K_8e^{-i\omega_2t}$$

where the K's are constants determined by the initial conditions. Of course, we physically expect those *complex*-valued exponentials to combine, via Euler's identity, into *real*-valued sinusoidal terms (actual, physical quantities that we can measure with instruments are *not* complex-valued!). Since both circuits, primary (circuit 1) and secondary (circuit 2), have zero current before the switch is closed (because of the inductors), we expect those currents to still be zero just after the switch is closed and so we argue that the K's must be such that

$$i_1(t) = A_1 \sin(\omega_1 t) + A_2 \sin(\omega_2 t) \tag{2.24}$$

and

$$i_2(t) = A_3 \sin(\omega_1 t) + A_4 \sin(\omega_2 t) \tag{2.25}$$

where the A's (obviously related to the K's) are determined by the initial conditions. Notice that (2.24) and (2.25) do, indeed, give $i_1(0+) = i_2(0+) = 0$.

The time domain equation of circuit 1 is (look back at (1.36)), with $R = 0$ and using the minus sign for M,

$$L_1 \frac{di_1}{dt} + \frac{1}{C_1} \int_0^t i_1(x)dx - M \frac{di_2}{dt} = V_0 \tag{2.26}$$

and, similarly for circuit 2 (look back at (1.37)),

$$L_2 \frac{di_2}{dt} + \frac{1}{C_2} \int_0^t i_2(x)dx - M \frac{di_1}{dt} = 0. \tag{2.27}$$

So, for $t = 0+$ we have

$$L_1 \frac{di_1}{dt} \Big|_{t=0+} - M \frac{di_2}{dt} \Big|_{t=0+} = V_0$$

and

$$L_2 \frac{di_2}{dt} \Big|_{t=0+} - M \frac{di_1}{dt} \Big|_{t=0+} = 0.$$

These two expressions are two equations in two unknowns, $\frac{di_1}{dt} \big|_{t=0+}$ and $\frac{di_2}{dt} \big|_{t=0+}$, and they are easily solved for with Cramer's rule. The system determinant is

$$D = \begin{vmatrix} L_1 & -M \\ -M & L_2 \end{vmatrix} = L_1 L_2 - M^2$$

and so

$$\frac{di_1}{dt}\Big|_{t=0+} = \frac{\begin{vmatrix} V_0 & -M \\ 0 & L_2 \end{vmatrix}}{D} = \frac{V_0 L_2}{D} = V_0 \frac{L_2}{L_1 L_2 - M^2} = X_1 \tag{2.28}$$

and

$$\frac{di_2}{dt}\Big|_{t=0+} = \frac{\begin{vmatrix} L_1 & V_0 \\ -M & 0 \end{vmatrix}}{D} = \frac{M V_0}{D} = V_0 \frac{M}{L_1 L_2 - M^2} = X_2. \tag{2.29}$$

You'll see, in just a bit, why X_1 and X_2 have been introduced (they will help keep the algebra from overwhelming us).

Next, let's differentiate (2.26) and (2.27) to get

$$L_1 \frac{d^2 i_1}{dt^2} + \frac{i_1(t)}{C_1} - M \frac{d^2 i_2}{dt^2} = 0$$

and

$$L_2 \frac{d^2 i_2}{dt^2} + \frac{i_2(t)}{C_2} - M \frac{d^2 i_1}{dt^2} = 0$$

which, since $i_1(0+) = i_2(0+) = 0$, says

$$L_1 \frac{d^2 i_1}{dt^2}\Big|_{t=0+} - M \frac{d^2 i_2}{dt^2}\Big|_{t=0+} = 0$$

and

$$L_2 \frac{d^2 i_2}{dt^2}\Big|_{t=0+} - M \frac{d^2 i_1}{dt^2}\Big|_{t=0+} = 0.$$

Again, we have two equations in two unknowns (that is, $\frac{d^2 i_1}{dt^2}\big|_{t=0+}$ and $\frac{d^2 i_2}{dt^2}\big|_{t=0+}$) and you should be able to see, *by inspection*, that the solutions are

$$\frac{d^2 i_1}{dt^2}\Big|_{t=0+} = 0 \tag{2.30}$$

and

$$\frac{d^2 i_2}{dt^2}\Big|_{t=0+} = 0. \tag{2.31}$$

Now, let's apply (2.28)/(2.29) and (2.30)/(2.31) to (2.24) and (2.25). Doing that for (2.28)/(2.29) gives us

$$A_1\omega_1 + A_2\omega_2 = V_0\frac{L_2}{L_1L_2 - M^2} = X_1 \tag{2.32}$$

and

$$A_3\omega_1 + A_4\omega_2 = V_0\frac{M}{L_1L_2 - M^2} = X_2 \tag{2.33}$$

But, alas, (2.30)/(2.31) gives us only $0 = 0$. Certainly true, but of no help in finding the *four* A's. For that we need four equations, and not just the two of (2.32) and (2.33). So, let's push on and differentiate yet again (look back at the equations immediately following (2.29)):

$$L_1\frac{d^3i_1}{dt^3} + \frac{1}{C_1}\frac{di_1}{dt} - M\frac{d^3i_2}{dt^3} = 0$$

and

$$L_2\frac{d^3i_2}{dt^3} + \frac{1}{C_2}\frac{di_2}{dt} - M\frac{d^3i_1}{dt^3} = 0$$

which, using (2.28) and (2.29) become (and now I'll use the X_1 and X_2 mentioned earlier) at $t = 0+$

$$L_1\frac{d^3i_1}{dt^3}\Big|_{t=0+} + \frac{1}{C_1}X_1 - M\frac{d^3i_2}{dt^3}\Big|_{t=0+} = 0$$

and

$$L_2\frac{d^3i_2}{dt^3}\Big|_{t=0+} + \frac{1}{C_2}X_2 - M\frac{d^3i_1}{dt^3}\Big|_{t=0+} = 0.$$

Thus, rearranging for Cramer's rule,

$$L_1\frac{d^3i_1}{dt^3}\Big|_{t=0+} - M\frac{d^3i_2}{dt^3}\Big|_{t=0+} = -\frac{1}{C_1}X_1$$

and

$$-M\frac{d^3i_1}{dt^3}\Big|_{t=0+} + L_2\frac{d^3i_2}{dt^3}\Big|_{t=0+} = -\frac{1}{C_2}X_2.$$

The solutions are (after a bit of easy algebra),

$$\frac{d^3 i_1}{dt^3}\Big|_{t=0+} = -V_0 \frac{C_2 L_2^2 + C_1 M^2}{C_1 C_2 \left(L_1 L_2 - M^2\right)^2} = Y_1 \tag{2.34}$$

and

$$\frac{d^3 i_2}{dt^3}\Big|_{t=0+} = -V_0 M \frac{C_1 L_1 + L_2 C_2}{C_1 C_2 \left(L_1 L_2 - M^2\right)^2} = Y_2. \tag{2.35}$$

Applying (2.34) and (2.35) to (2.24) and (2.25), we get

$$-A_1 \omega_1^3 - A_2 \omega_2^3 = Y_1 \tag{2.36}$$

and

$$-A_3 \omega_1^3 - A_4 \omega_2^3 = Y_2. \tag{2.37}$$

With (2.32), (2.33), (2.36), and (2.37), we at last have four equations for the four A's. That is,

(a) $A_1\omega_1 + A_2\omega_2 = X_1$
(b) $A_3\omega_1 + A_4\omega_2 = X_2$
(c) $-A_1\omega_1^3 - A_2\omega_2^3 = Y_1$
(d) $-A_3\omega_1^3 - A_4\omega_2^3 = Y_2$.

From the structure of these equations we can obviously solve (a) and (c) together to get A_1 and A_2, and (b) and (d) together to get A_3 and A_4. Doing that, we arrive at

$$A_1 = \frac{-X_1\omega_2^3 - Y_1\omega_2}{\omega_2\omega_1^3 - \omega_1\omega_2^3}, A_2 = \frac{X_1\omega_1^3 + Y_1\omega_1}{\omega_2\omega_1^3 - \omega_1\omega_2^3}, A_3 = \frac{-X_2\omega_2^3 - Y_2\omega_2}{\omega_2\omega_1^3 - \omega_1\omega_2^3},$$
$$A_4 = \frac{X_2\omega_1^3 + Y_2\omega_1}{\omega_2\omega_1^3 - \omega_1\omega_2^3}.$$

The ω's are given by (2.22) and (2.23), the X's by (2.32) and (2.33), and the Y's by (2.34) and (2.35). That's a *lot* of number-crunching, yes, but for a computer it's all just wood-chips to pulverize, and so it's duck soup to generate curves for $i_1(t)$ and $i_2(t)$ from (2.24) and (2.25). For example, Fig. 2.10 shows such curves for the values $V_0 = 1$ volt, $L_1 = 93\ mH$, $C_1 = 150\ \mu F$, $L_2 = 11\ mH$, and $C_2 = 168\ \mu F$, for the two cases of 'loose' magnetic coupling ($M = 0.1\sqrt{L_1 L_2}$) and 'tight' magnetic coupling ($M = 0.7\sqrt{L_1 L_2}$).

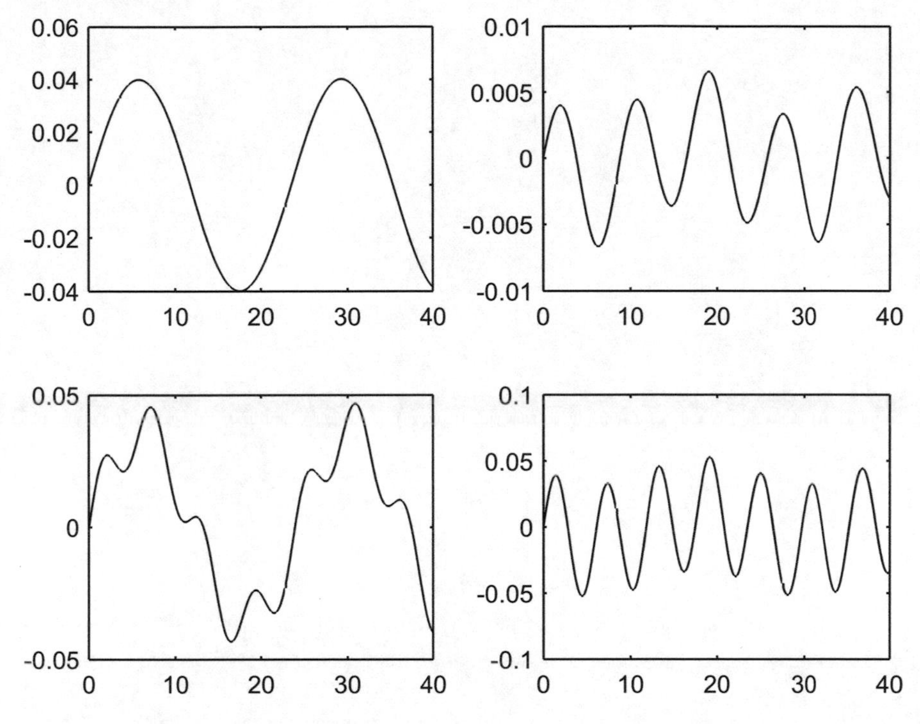

Fig. 2.10 The primary (left) and secondary (right) currents for 'loose' coupling (top) and 'tight' coupling (bottom). The current is in amperes and time is in milliseconds

Problems

2.1 Show that (2.20) has the property given in (2.11).

Chapter 3
The Laplace Transform

3.1 The Transform, and Why It's Useful

As you've seen in the first two chapters, the time domain analysis of electrical circuits results in equations containing time derivatives. Often a *lot* of derivatives. This doesn't necessarily make progress impossible, but time derivatives do add significantly to the horrors of calculation. *The Laplace transformation allows us to get rid of differentiations*.

The transform can do this because it has the wonderful property of converting the operation of differentiation into the far simpler one of multiplication. That is, it *transforms* a *differential* equation into an *algebraic* equation. This process is analogous to how logarithms transform multiplication into the simpler operation of addition (because $\log(xy) = \log(x) + \log(y)$).

If $f(t)$ is a time function, then we write the Laplace transform of $f(t)$ as $F(s)$, where[1]

$$F(s) = \mathcal{L}\{f(t)\} = \int_0^\infty f(t)e^{-st}dt \tag{3.1}$$

where s is a complex quantity written as $s = \sigma + i\omega$. The role of σ (which always has a finite *positive* value) is to make the decaying, oscillatory $e^{-st} = e^{-\sigma t}e^{-i\omega t}$ factor go to zero fast enough that the integral in (3.1) actually exists. It turns out, you should understand, that there *are* some easy-to-write time functions which blow-up so fast (as $t \to \infty$) that it is impossible to find such a σ, functions like t^t and e^{t^2}. This isn't of any practical concern, however, because such explosive time functions simply don't

[1]When I took my first course in ordinary differential equations at Stanford (Math 130, Autumn 1959), and the transform was introduced, I stuck my hand up in class that day and innocently asked the instructor 'How did Laplace know to do this?' The instructor just smiled and said 'Because he was a really smart guy.' We all laughed, but he really should have said 'Because he had read Euler' (see note 15 in the Preface). Okay, you say, so how did *Euler* know to do this? What can I say: because he was a really smart guy.

P. J. Nahin, *Transients for Electrical Engineers*,
https://doi.org/10.1007/978-3-319-77598-2_3

occur in actual circuits. Certainly all of the time functions of interest in the transient analyses we do here will have a Laplace transform. Do notice, carefully, that (3.1) ignores what $f(t)$ was 'doing' for $t < 0$. It is only the behavior of $f(t)$ for $t \geq 0$ that determines $F(s)$. I'll say more on this issue in the next section.

The Laplace transform is a *unique* operation, which means different $f(t)$'s have different $F(s)$'s. The two functions, $f(t)$ and $F(s)$, are often written, in fact, as

$$f(t) \leftrightarrow F(s)$$

and called a *transform pair*. The double-headed arrow indicates there is a one-to-one correspondence between a time function and its transform.

What makes the definition of (3.1) useful is the following result:

$$\mathcal{L}\left\{\frac{df}{dt}\right\} = \int_0^\infty e^{-st}\frac{df}{dt}dt = sF(s) - f(0+) \tag{3.2}$$

where $f(0+)$ is the value of $f(t)$ just after some switching event at $t = 0$ for $f(t)$. When $f(t)$ is continuous at $t = 0$ then, of course, $f(0) = f(0+)$. We write $f(0+)$ — the value of $f(t)$ as $t \to 0$ *from positive values* — instead of $f(0)$, in anticipation of the possibility $f(t)$ might have a discontinuity at $t = 0$. That is, differentiation in the time domain becomes multiplication by s in the transform domain. In addition, you see in (3.2) how the initial condition $f(0+)$ comes into play and, while you might think this is a complication, *it is not*. This, in fact, is precisely what Heaviside's operational calculus failed to do (and, as a result, Heaviside always had to assume $f(0+) = 0$, as you'll recall was mentioned in Carter's book: see note 11 in the Preface).

The proof of (3.2) follows easily by doing the integral by-parts. That is, in the traditional notation used by all calculus textbooks,[2]

$$\int_0^\infty u\,dv = (uv)\,|_0^\infty - \int_0^\infty v\,du \tag{3.3}$$

where we write $dv = \frac{df}{dt}dt = df$ (and so $v = f$), and $u = e^{-st}$ (and so $du = -se^{-st}dt$). Thus,

$$\mathcal{L}\left\{\frac{df}{dt}\right\} = \{f(t)e^{-st}\}\,|_0^\infty - \int_0^\infty f(t)(-se^{-st}dt)$$
$$= \{f(t)e^{-st}\}\,|_0^\infty + s\int_0^\infty f(t)e^{-st}dt.$$

[2]This formula follows immediately from the integration of the freshman calculus formula for the differential of a product: $d(uv) = udv + vdu$.

That is,

$$\mathcal{L}\left\{\frac{df}{dt}\right\} = \{f(t)e^{-st}\}\Big|_0^\infty + sF(s). \tag{3.4}$$

We next argue that the first term on the right in (3.4) is

$$\{f(t)e^{-st}\}\Big|_{0+}^\infty = -f(0+)$$

because, for the transform to exist, we must have σ sufficiently positive that

$$\lim_{t\to\infty} f(t)e^{-(\sigma+i\omega)t} = 0.$$

This requires only that $f(\infty)$ not be 'too large.' Certainly, assuming $f(\infty)$ is finite works, and this is a pretty safe assumption for the voltages and currents in any circuit we can actually build. Thus, (3.4) becomes (3.2). You can repeat these same arguments, over and over, to show that

$$\mathcal{L}\left\{\frac{d^2f}{dt^2}\right\} = s^2F(s) - sf(0+) - f'(0+) \tag{3.5}$$

$$\mathcal{L}\left\{\frac{d^3f}{dt^3}\right\} = s^3F(s) - s^2f(0+) - sf'(0+) - f''(0+), \tag{3.6}$$

and so on, where $f'(0+)$ and $f''(0+)$ are the values at $t = 0+$ of the first and second derivatives of $f(t)$.

If multiplication by s is associated with time differentiation, then a natural question to ask is 'what time operation is *division* by s associated with in the time domain?' Could it be, just as a wild guess, *integration*? The answer is yes, and here's how to see that. Given that $F(s)$ is the Laplace transform of $f(t)$, let's develop the transform of the time function (notice the upper limit)

$$\int_0^t f(x)dx$$

where x is simply a so-called 'dummy variable of integration.'[3] That is, let's calculate

$$\mathcal{L}\left\{\int_0^t f(x)dx\right\} = \int_0^\infty \left\{\int_0^t f(x)dx\right\}e^{-st}dt.$$

[3] You'll occasionally see in textbooks such things as $\int_0^t f(t)dt$ which is one of those expressions that falls into the category of 'everybody knows what is meant even though it is badly stated.' I say that because such an expression is saying t varies from 0 to t, which is meaningless. The same symbol cannot represent both an integration *variable* and an integration *limit*.

Integrating by-parts, define[4]

$$u = \int_0^t f(x)dx \text{ and so } du = f(t)\frac{dt}{dt}dt = f(t)dt$$

and

$$dv = e^{-st}dt \text{ and so } v = -\frac{1}{s}e^{-st}.$$

Then,

$$\mathcal{L}\left\{\int_0^t f(x)dx\right\} = \left\{-\frac{1}{s}e^{-st}\int_0^t f(x)dx\right\}\Big|_0^\infty - \int_0^\infty -\frac{1}{s}e^{-st}f(t)dt$$

and so

$$\mathcal{L}\left\{\int_0^t f(x)dx\right\} = \frac{1}{s}\int_0^\infty e^{-st}f(t)dt = \frac{F(s)}{s} \tag{3.7}$$

because $e^{-st}\int_0^t f(x)dx$ vanishes at both $t = 0$ (*if* $\int_0^0 f(x)dx = 0$) and at $t = \infty$ (*if* $\int_0^t f(x)dx$ doesn't 'blow-up' too fast as $t \to \infty$). So, as we might have anticipated from (3.2), integrating in the time domain results in dividing (by s) in the transform domain.

We can find another useful result by, instead of integrating $f(t)$ as in (3.7), differentiating (with respect to s) the Laplace transform definition in (3.1):

$$\frac{dF}{ds} = \frac{d}{ds}\int_0^\infty f(t)e^{-st}dt = \int_0^\infty f(t)(-t)e^{-st}dt$$

or,

$$\int_0^\infty tf(t)e^{-st}dt = -\frac{dF}{ds}.$$

That is, given the pair $f(t) \leftrightarrow F(s)$, then another pair is

$$tf(t) \leftrightarrow -\frac{dF}{ds}. \tag{3.8}$$

[4]To calculate $\frac{du}{dt}$, use Leibniz's formula (see note 3 in Chap. 1).

Finally, a mirror-image to (3.8) is: given the pair $f(t) \leftrightarrow F(s)$, then

$$\frac{f(t)}{t} \leftrightarrow \int_s^\infty F(x)dx, \tag{3.9}$$

which we can prove by simply evaluating the integral. Thus,

$$\int_s^\infty F(x)dx = \int_s^\infty \left\{ \int_0^\infty f(t)e^{-xt}dt \right\} dx$$

or, assuming (as engineers do at the drop of a hat) that we can reverse the order of integration,

$$\begin{aligned}\int_s^\infty F(x)dx &= \int_0^\infty f(t)\left\{ \int_s^\infty e^{-xt}dx \right\} dt = \int_0^\infty f(t)\left\{ -\frac{e^{-xt}}{t} \right\} |_s^\infty dt \\ &= \int_0^\infty \frac{f(t)}{t} e^{-st}dt \\ &= \mathcal{L}\left\{ \frac{f(t)}{t} \right\}.\end{aligned}$$

3.2 The Step, Exponential, and Sinusoid Functions of Time

In the rest of this chapter we'll develop the Laplace transform further, finding both some additional general theorems that will prove useful and the transforms of some specific, common time functions. One such time function plays a role so central to transient analysis, however, that there can be no further delay in introducing it. This is the *unit step function*, often called the *Heaviside step* because Oliver Heaviside made such enormous use of it in his operational calculus; it is the mathematical description of a time function that is suddenly switched-on at time $t = 0$. Written as $u(t)$, it is

$$u(t) = \begin{matrix} 1, & t > 0 \\ 0, & t < 0 \end{matrix}. \tag{3.10}$$

Note, *carefully*, that $u(0)$ is *not* defined. An often-made 'natural' proposal is that $u(0) = \frac{1}{2}$ and, with a somewhat more mathematically elevated level than used in this book, we can actually lend some strength to that idea.[5] The specific value of $u(0)$ is not important here, but we will revisit the nature of $u(t)$ — and specifically the value of $u(0)$ — in Problem 3.13.

[5] See, for example, my book *Dr. Euler's Fabulous Formula*, Princeton 2017, pp. 209–211.

The Laplace transform of $u(t)$ is easy to calculate:

$$\mathcal{L}\{u(t)\} = \int_0^\infty u(t)e^{-st}dt = \int_0^\infty e^{-st}dt = \left\{-\frac{1}{s}e^{-st}\right\}\Big|_0^\infty$$

and so

$$\mathcal{L}\{u(t)\} = \frac{1}{s}. \tag{3.11}$$

The exponential function e^{-at}, where a is some constant, is just about as easy to transform:

$$\mathcal{L}\{e^{-at}\} = \int_0^\infty e^{-at}e^{-st}dt = \int_0^\infty e^{-(s+a)t}dt = \left\{-\frac{1}{s+a}e^{-(s+a)t}\right\}\Big|_0^\infty$$

or,

$$\mathcal{L}\{e^{-at}\} = \frac{1}{s+a}. \tag{3.12}$$

Notice that if $a = 0$ the exponential time function reduces to the unit step in time, and the transform in (3.12) does indeed become the transform in (3.11).

A generalization of (3.12) is the theorem

$$\mathcal{L}\{e^{-at}f(t)\} = F(s+a) \tag{3.13}$$

which follows from

$$\mathcal{L}\{e^{-at}f(t)\} = \int_0^\infty e^{-at}f(t)e^{-st}dt = \int_0^\infty f(t)e^{-(s+a)t}dt$$

which, if we change variable to $p = s + a$, becomes

$$\mathcal{L}\{e^{-at}f(t)\} = \int_0^\infty f(t)e^{-pt}dt = F(p) = F(s+a).$$

The Laplace transform pair $f(t) \leftrightarrow F(s)$ is independent of what $f(t)$ may have been doing for $t < 0$. That earlier behavior may influence the values of $f(0)$, $f'(0)$ and so on, but that's the *only* impact that earlier behavior has on $F(s)$. It is therefore understood that when we calculate $F(s)$ for a particular $f(t)$, what is actually being calculated is the transform of $f(t)u(t)$ and that the transform pair is $f(t)u(t) \leftrightarrow F(s)$.

Another highly useful result in transform theory, one involving the step that looks a bit like (3.13) but is actually quite different, is the so-called *shifting theorem*:

$$\mathcal{L}\{f(t - t_0)u(t - t_0)\} = e^{-st_0}F(s). \tag{3.14}$$

That is, if we shift the time function $f(t)u(t)$ by t_0 ($t_o > 0$ corresponds to a *delay*)[6] then we multiply the transform of $f(t)u(t)$ by e^{-st_0}. This follows easily:

$$\mathcal{L}\{f(t - t_0)u(t - t_0)\} = \int_0^\infty f(t - t_0)u(t - t_0)e^{-st}dt = \int_{t_0}^\infty f(t - t_0)e^{-st}dt$$

because $u(t - t_0) = 0$ for $t < t_0$. Now, let $p = t - t_0$ (and so $t = p + t_0$ and $dt = dp$). Then our integral becomes (p is, of course, simply a dummy variable of integration)

$$\int_0^\infty f(p)e^{-s(p+t_0)}dp = e^{-st_0}\int_0^\infty f(p)e^{-sp}dp = e^{-st_0}F(s).$$

To compute the transforms of the sinusoidal time functions $\sin(\omega_0 t)$ and $\cos(\omega_0 t)$, we can use Euler's identity, along with (3.12), as follows. As shown in Appendix 1,

$$\cos(\omega_0 t) = \frac{e^{i\omega_0 t} + e^{-i\omega_0 t}}{2}.$$

Thus, associating s with $i\omega_0$ or with $-i\omega_0$, and using (3.12), we see that[7]

$$\mathcal{L}\{\cos(\omega_0 t)\} = \frac{1}{2}\left(\frac{1}{s - i\omega_0} + \frac{1}{s + i\omega_0}\right) = \frac{1}{2}\left(\frac{s + i\omega_0 + s - i\omega_0}{s^2 + \omega_0^2}\right)$$

and so

$$\mathcal{L}\{\cos(\omega_0 t)\} = \frac{s}{s^2 + \omega_0^2}. \tag{3.15}$$

In the same way, since

$$\sin(\omega_0 t) = \frac{e^{i\omega_0 t} - e^{-i\omega_0 t}}{2i}$$

then

$$\mathcal{L}\{\sin(\omega_0 t)\} = \frac{1}{2i}\left(\frac{1}{s - i\omega_0} - \frac{1}{s + i\omega_0}\right) = \frac{1}{2i}\left(\frac{s + i\omega_0 - s + i\omega_0}{s^2 + \omega_0^2}\right)$$

[6]Notice that $f(t - t_0)u(t - t_0) = 0$ for $t < t_o$, while $f(t - t_0)$ by itself is *not* in general identically zero.

[7]Also used is $\mathcal{L}\{f(t) + g(t)\} = F(s) + G(s)$, which is a statement that the integral of a sum is the sum of the integrals.

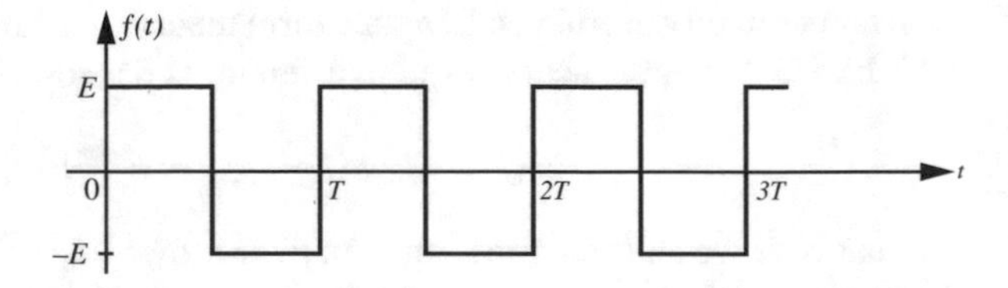

Fig. 3.1 A square-wave

and so

$$\mathcal{L}\{\sin(\omega_0 t)\} = \frac{\omega_0}{s^2 + \omega_0^2}. \tag{3.16}$$

To state the obvious, $\sin(\omega_0 t)$ and $\cos(\omega_0 t)$ are particular examples of the class of functions called *periodic*. That is, of functions for which there is a value T (called the *period*, which is the duration of a *cycle*) such that

$$f(t) = f(t + T).$$

The sinusoids have long been associated with electrical engineering as they are available out of every household wall plug in the civilized world,[8] but there *are* other important possibilities. A *non*-sinusoidal example (a *square-wave*) is shown in Fig. 3.1.

We can express the Laplace transform of such a function in terms of the Laplace transform of the *first cycle alone*. If we call that first cycle $f_1(t)$, and its transform $F_1(s)$, then

$$F_1(s) = \int_0^T f_1(t)e^{-st}dt$$

and the transform $F(s)$ of the complete, periodic $f(t)$ is

$$F(s) = \frac{F_1(s)}{1 - e^{-Ts}}. \tag{3.17}$$

This is easily proved as follows, with the aid of the step function. Observing that $f(t)$ can be written as $f_1(t)$ plus an endless number of increasingly delayed versions of that first cycle; that is, as the first cycle, plus the second cycle, plus the third cycle, plus . . ., we have

[8]Have you ever wondered why the very first water-powered electrical generators of alternating voltage/current were *sinusoidal* generators, and not some other function (like square waves)? It was all a fortuitous accident: rotate a coil of wire at uniform speed in the uniform magnetic field of a permanent magnet (as would happen, for example, if the coil is being rotated by the shaft of a paddlewheel of a hydro-power station next to a river flowing at constant speed) and the resulting voltage and current are *automatically* sinusoidal, along with all the nice mathematical properties of sinusoids that nobody was thinking about until much later.

$$f(t) = f_1(t) + f_2(t) + f_3(t) + \ldots.$$

Recognizing that $f_2(t)$ is $f_1(t)$ shifted by T, that $f_3(t)$ is $f_1(t)$ shifted by $2T$, and so on, we have

$$f(t) = f_1(t) + f_1(t-T)u(t-T) + f_1(t-2T)u(t-2T) + \ldots.$$

Now, recalling the shifting theorem of (3.14), we see that

$$F(s) = F_1(s) + e^{-sT}F_1(s) + e^{-s2T}F_1(s) + \ldots$$

or,

$$F(s) = F_1(s)\left[1 + e^{-sT} + e^{-2sT} + \ldots\right].$$

The quantity in the square brackets is an easily summed geometric series and, as claimed,

$$F(s) = \frac{F_1(s)}{1 - e^{-Ts}}.$$

As an example of (3.17), let's calculate the transform of the square-wave in Fig. 3.1. The first cycle is

$$f_1(t) = \begin{matrix} E, & 0 < t < \frac{T}{2} \\ -E, & \frac{T}{2} < t < T \end{matrix}$$

and so

$$F_1(s) = \int_0^{T/2} Ee^{-st}dt - \int_{T/2}^{T} Ee^{-st}dt = E\left[\left(-\frac{e^{-st}}{s}\right)\Big|_0^{T/2} - \left(-\frac{e^{-st}}{s}\right)\Big|_{T/2}^{T}\right]$$

$$= E\left[\frac{1 - e^{-s\frac{T}{2}}}{s} - \frac{e^{-s\frac{T}{2}} - e^{-sT}}{s}\right] = E\left[\frac{1 - 2e^{-s\frac{T}{2}} + e^{-sT}}{s}\right] = E\frac{\left(1 - e^{-s\frac{T}{2}}\right)^2}{s}.$$

Thus, by (3.17), we have the Laplace transform of the square-wave in Fig. 3.1 as

$$E\frac{\left(1 - e^{-s\frac{T}{2}}\right)^2}{s(1 - e^{-Ts})} = E\frac{\left(1 - e^{-s\frac{T}{2}}\right)\left(1 - e^{-s\frac{T}{2}}\right)}{s\left(1 - e^{-s\frac{T}{2}}\right)\left(1 + e^{-s\frac{T}{2}}\right)} = \frac{E}{s}\frac{1 - e^{-s\frac{T}{2}}}{1 + e^{-s\frac{T}{2}}} = \frac{E}{s}\frac{e^{-\frac{T}{4}s}\left(e^{\frac{T}{4}s} - e^{-\frac{T}{4}s}\right)}{e^{-\frac{T}{4}s}\left(e^{\frac{T}{4}s} + e^{-\frac{T}{4}s}\right)}$$

$$= \frac{E}{s}\frac{e^{\frac{T}{4}s} - e^{-\frac{T}{4}s}}{e^{\frac{T}{4}s} + e^{-\frac{T}{4}s}}.$$

Recalling the definitions of the hyperbolic sine and cosine,

$$sinh(x) = \frac{e^x - e^{-x}}{2}$$

and

$$cosh(x) = \frac{e^x + e^{-x}}{2},$$

we see that the transform of the square-wave in Fig. 3.1 is a hyperbolic tangent:

$$\frac{E}{s}\frac{sinh\left(\frac{T}{4}s\right)}{cosh\left(\frac{T}{4}s\right)} = \frac{E}{s}tanh\left(\frac{T}{4}s\right).$$

Now, as long as we are proving general theorems, let me end this section with two more. To repeat (3.2) and then adding a bit more,

$$\mathcal{L}\left(\frac{df}{dt}\right) = sF(s) - f(0+) = \int_0^\infty \frac{df}{dt}e^{-st}dt = \int_{f(0)}^{f(\infty)} e^{-st}df.$$

So, taking the limit as $s \to 0$, we have (assuming $f(t)$ is continuous and so $f(0) = f(0+)$)

$$\lim_{s\to 0}sF(s) - f(0+) = \lim_{s\to 0}\int_{f(0+)}^{f(\infty)} e^{-st}df = \int_{f(0+)}^{f(\infty)} \lim_{s\to 0} e^{-st}df = \int_{f(0+)}^{f(\infty)} df$$
$$= f(\infty) - f(0+),$$

assuming (of course) that it's legitimate to interchange the order of the limiting and the integration operations (and being engineers, we *will* so assume). Then, cancelling the two $f(0+)$ terms, we have what is called the *final value theorem*:

$$\lim_{t\to\infty} f(t) = f(\infty) = \lim_{s\to 0}sF(s). \tag{3.18}$$

An important exception to keep in mind about the final value theorem is that it does *not* apply to periodic functions, simply because such functions *don't have* 'final' values!

As you might expect (or at least hope) there is an *initial value theorem*, too:

$$\lim_{t\to 0} f(t) = f(0+) = \lim_{s\to\infty}sF(s). \tag{3.19}$$

The proof goes through almost the same way as did the demonstration of the final value theorem. We write, as before,

$$\mathcal{L}\left(\frac{df}{dt}\right) = \int_0^\infty \frac{df}{dt} e^{-st} dt = sF(s) - f(0+)$$

but now, when we let $s \to \infty$ in the integral, we see that $\lim_{s\to\infty} e^{-st} = 0$, the integral vanishes, and so

$$\lim_{s\to\infty} sF(s) - f(0+) = 0$$

and we are done.

3.3 Two Examples of the Transform in Action

Now, at last, with no further delay, let's address the question that has surely been ticking away in your brain: how do we *use* the Laplace transform to answer questions about transients in electrical circuits? In the next chapter I'll show you some detailed examples of doing just that but, for right now, here are a couple of simpler yet still quite illuminating examples that are, if attacked strictly in the time domain, not at all trivial. In the s-domain, however, matters are significantly easier.

For the first example, consider the circuit of Fig. 3.2, which shows the application (at time $t = 0$) of the voltage source $sin(\omega_0 t + \theta)$ to a resistor and capacitor connected in series. The problem is to show that, if the capacitor has no initial charge, then there is a value to the phase angle θ (determined by the values of R, C, and ω_0) that results in a current that *has no transient*. That is, we are to show that there is a θ such that when the switch is closed at $t = 0$ the current $i(t)$ instantly takes-on its *steady-state* form as a pure, *undamped* sinusoidal oscillation.

Writing Kirchhoff's voltage law around the loop,

$$v(t) = i(t)R + \frac{1}{C}\int_0^t i(u)du$$

or, differentiating,

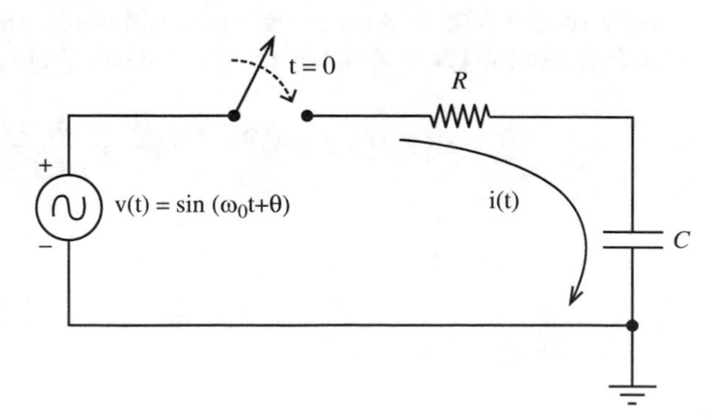

Fig. 3.2 What is θ for a non-transient $i(t)$?

$$\frac{dv}{dt} = R\frac{di}{dt} + \frac{1}{C}i$$

and so, transforming,

$$sV(s) - v(0+) = R[sI(s) - i(0+)] + \frac{1}{C}I(s). \tag{3.20}$$

Now, since the voltage drop across the capacitor is initially zero, and since that drop can't change instantly, we see that

$$i(0+) = \frac{v(0+)}{R}.$$

So,

$$sV(s) - v(0+) = RsI(s) - Ri(0+) + \frac{1}{C}I(s)$$

or,

$$sV(s) - v(0+) = RsI(s) - R\frac{v(0+)}{R} + \frac{1}{C}I(s)$$

or,

$$sV(s) = RsI(s) + \frac{1}{C}I(s) = I(s)\left[Rs + \frac{1}{C}\right]$$

and so (3.20) becomes

$$I(s) = \frac{sV(s)}{Rs + \frac{1}{C}} = \frac{1}{R}\left(\frac{sV(s)}{s + \frac{1}{RC}}\right). \tag{3.21}$$

Our plan of attack is now straightforward. First, we'll insert $V(s)$ into (3.21), and then see if the resulting expression for $I(s)$ 'looks-like' any of the transforms we've already calculated (if so, those transforms will give us the time behavior $i(t)$ that goes with $I(s)$). For the given $V(s)$, it is not difficult (see Problem 3.1) to show that

$$\mathcal{L}\{\sin(\omega_0 t + \theta)\} = \frac{s\sin(\theta) + \omega_0\cos(\theta)}{s^2 + \omega_0^2}, \tag{3.22}$$

and you see that (3.22) reduces to the special cases we've already worked out for $\theta = 0$ and $\theta = \frac{\pi}{2}$ (look back at (3.15) and (3.16)). So,

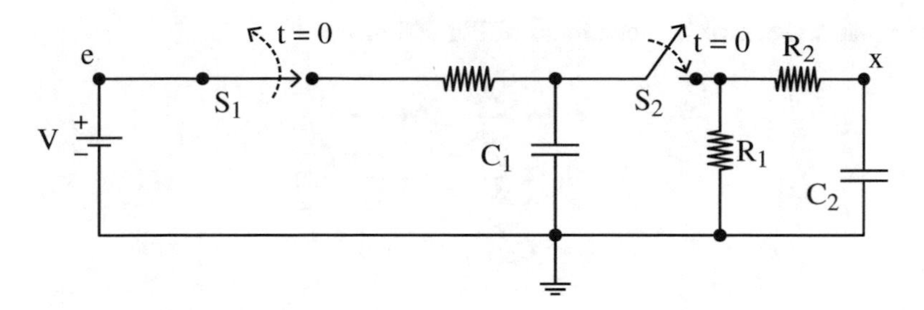

Fig. 3.3 What is $x(t)$ for $t \geq 0$?

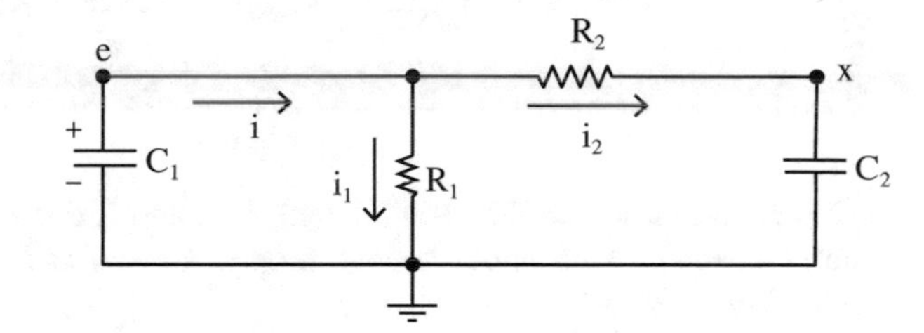

Fig. 3.4 The circuit of Fig. 3.4 for $t > 0$

$$I(s) = \frac{1}{R} \frac{s\left[\frac{s\sin(\theta)+\omega_0\cos(\theta)}{s^2+\omega_0^2}\right]}{s+\frac{1}{RC}} = \frac{\sin(\theta)}{R} \frac{s\left[s+\frac{\omega_0\cos(\theta)}{\sin(\theta)}\right]}{\left(s^2+\omega_0^2\right)\left(s+\frac{1}{RC}\right)}.$$

Notice that if

$$\frac{\omega_0\cos(\theta)}{\sin(\theta)} = \frac{1}{RC} = \frac{\omega_0}{\tan(\theta)}$$

then common factors in the numerator and denominator of $I(s)$ cancel and the transform collapses to

$$I(s) = \frac{\sin(\theta)}{R} \frac{s}{\left(s^2+\omega_0^2\right)}$$

which is the transform of a pure sinusoid time function; in other words, there is no transient in $i(t)$ if $\theta = tan^{-1}(\omega_0 RC)$.

For a second example of the use of the Laplace transform, consider the circuit of Fig. 3.3, where switch S_2 is open and switch S_1 has been closed for a long time. Thus, capacitor C_1 has charged to V volts, while capacitor C_2 has a voltage drop of zero. Then, at time $t = 0$, S_1 is opened and S_2 is closed. What is $x(t)$, the voltage drop across C_2, for $t \geq 0$? For $t > 0$, the circuit looks as shown in Fig. 3.4.

The equations describing the circuit of Fig. 3.4 are:

$$i = -C_1\frac{de}{dt}, e(0+) = V$$
$$i_1 = \frac{e}{R_1}, i_2 = \frac{e-x}{R_2}$$
$$i_2 = C_2\frac{dx}{dt}, x(0+) = 0.$$

Transforming, these equations become

$$I = -C_1[sE - e(0+)] = -C_1sE + C_1V$$
$$I_1 = \frac{E}{R_1}, \quad I_2 = \frac{E-X}{R_2}$$
$$I_2 = C_2sX.$$

Note, *carefully*, that the capital letters I, E, and X denote transforms, while V is simply a *number* (the initial voltage drop across C_1 at $t = 0$). Since Kirchhoff's current law says.

$$I = I_1 + I_2,$$

we have

$$-C_1sE + C_1V = \frac{E}{R_1} + \frac{E-X}{R_2}, \tag{3.23}$$

while equating the two expressions for I_2 says

$$\frac{E-X}{R_2} = C_2sX.$$

Thus,

$$E - X = R_2C_2sX$$

and so

$$E = X(1 + R_2C_2s). \tag{3.24}$$

Inserting (3.24) into (3.23), we arrive at

$$-C_1sX(1 + R_2C_2s) + C_1V = \frac{X(1 + R_2C_2s)}{R_1} + XC_2s$$

or,

$$-R_1C_1sX(1 + R_2C_2s) + R_1C_1V = X(1 + R_2C_2s) + XR_1C_2s$$

or,

$$X(1 + R_2C_2s) + XR_1C_2s + R_1C_1sX(1 + R_2C_2s) = C_1VR_1$$

or,

$$X\left[1 + R_2C_2s + R_1C_2s + R_1C_1s + R_1C_1R_2C_2s^2\right] = C_1VR_1$$

or,

$$\frac{X(s)}{V} = \frac{C_1R_1}{1 + (R_2C_2 + R_1C_2 + R_1C_1)s + R_1C_1R_2C_2s^2}. \tag{3.25}$$

This is getting a bit clunky to handle symbolically, so let's continue with some specific numbers: in particular,

$$\begin{aligned} C_1 &= 0.02\ \mu fd = 2 \times 10^{-2}\ \mu fd \\ C_2 &= 0.001\ \mu fd = 10^{-3}\ \mu fd \\ R_1 &= 5,000\ ohms = 5 \times 10^3\ ohms \\ R_2 &= 1,000\ ohms = 10^3\ ohms \end{aligned}$$

where I've expressed capacitance in units of microfarads and resistance in units of ohms to get time in units of microseconds. Then (3.25) becomes

$$\frac{X(s)}{V} = \frac{5 \times 10^3 \times 2 \times 10^{-2}}{5 \times 10^3 \times 10^3 \times 2 \times 10^{-2} \times 10^{-3}s^2 + (10^3 \times 10^{-3} + 5 \times 10^3 \times 10^{-3} + 5 \times 10^3 \times 2 \times 10^{-2})s + 1}$$

or, with some simplification,

$$\frac{X(s)}{V} = \frac{1}{s^2 + 1.06s + 0.01}. \tag{3.26}$$

Remember, $X = X(s)$ is a *transform*, while V is simply a number.

To invert $X(s)$ back to $x(t)$, we need to get $X(s)$ into a form resembling one (or more) of the standard transforms we've already worked-out. This process will be our first example of what is called a *partial fraction expansion*. In particular, since we know the quadratic denominator can be expressed as

$$s^2 + 1.06s + 0.01 = (s + \alpha_1)(s + \alpha_2),$$

then

$$\frac{x(t)}{V} = \frac{1}{(s + \alpha_1)(s + \alpha_2)} = \frac{A}{s + \alpha_1} + \frac{B}{s + \alpha_2}$$

where A and B, and α_1 and α_2, are four constants (to be determined, soon) and so, using (3.12), we return to the time domain and write

$$\frac{x(t)}{V} = Ae^{-\alpha_1 t} + Be^{-\alpha_2 t}, t \geq 0. \tag{3.27}$$

Almost done! All we have left to do is the calculation of the constants α_1, α_2, A, and B. (We know, *physically*, before doing any calculation, that α_1 and α_2 must *both* be *positive* numbers because $x(t)$ will *decay* with increasing time.)

For α_1 and α_2, notice that

$$s^2 + 1.06s + 0.01 = s^2 + (\alpha_1 + \alpha_2)s + \alpha_1\alpha_2$$

and so

$$\alpha_1 + \alpha_2 = 1.06$$

and

$$\alpha_1\alpha_2 = 0.01.$$

These two equations are easily solved to give $\alpha_1 = 0.00952$ and $\alpha_2 = 1.05$. So,

$$\frac{x(t)}{V} = \frac{A}{s + 0.00952} + \frac{B}{s + 1.05} = \frac{1}{(s + 0.00952)(s + 1.05)}. \tag{3.28}$$

Thus, multiplying through (3.28) by (s + 0.00952),

$$A + (s + 0.00952)\frac{B}{s + 1.05} = \frac{1}{s + 1.05}$$

or, setting $s = -0.00952$,

$$A = \frac{1}{s + 1.05} \mid_{s=-0.00952} = 0.961.$$

Similarly, multiplying through (3.28) by (s + 1.05),

$$(s + 1.05)\frac{A}{s + 0.00952} + B = \frac{1}{s + 0.00952}$$

or, setting $s = -1.05$,

$$B = \frac{1}{s + 0.00952} \mid_{s=-1.05} = -0.961.$$

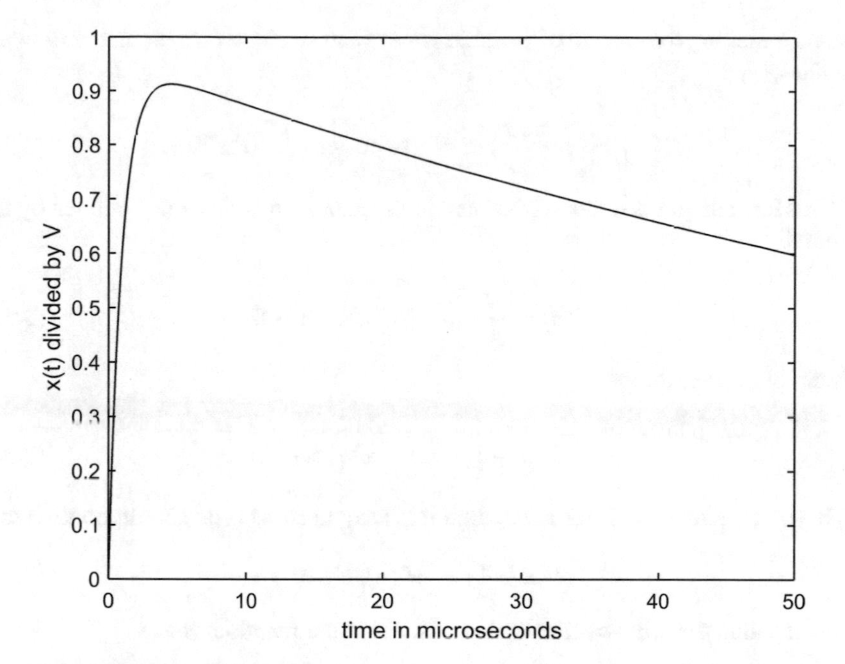

Fig. 3.5 The answer to the question of Fig. 3.3

Therefore, (3.27) becomes

$$\frac{x(t)}{V} = 0.961\left(e^{-0.00952t} - e^{-1.05t}\right), t \geq 0. \tag{3.29}$$

Figure 3.5 shows what (3.29) looks like: a *very* rapid rise to a peak voltage, followed by a much more gradual decline to zero.

3.4 Powers of Time

In this section we'll compute the Laplace transforms of $f(t) = t^n$, where n is any non-negative integer. (The case of $n = 0$ will reproduce our earlier result for $f(t) = u(t)$, the unit step function, which will be a partial check on our calculations.) The case of $n = 1$, in particular, represents the time function of a linearly increasing *ramp*, a function that commonly occurs in electrical engineering. And, perhaps surprisingly, we'll get 'bonus' results for the curious cases of $n = \pm\frac{1}{2}$; that is, we'll also get the Laplace transforms for the time functions $\frac{1}{\sqrt{t}}$ and $\sqrt{t}$.

We start by writing the formal answer as

$$\mathcal{L}\{t^n\} = \int_0^\infty t^n e^{-st} dt \tag{3.30}$$

and then making the change of variable $st = x$ (and so $t = x/s$ which means $dt = dx/s$). Thus,

$$\mathcal{L}\{t^n\} = \int_0^\infty \left(\frac{x}{s}\right)^n e^{-s\frac{x}{s}} \frac{1}{s} dx = \frac{1}{s^{n+1}} \int_0^\infty x^n e^{-x} dx. \tag{3.31}$$

This last integral is intimately related to the *gamma function* $\Gamma(n)$, defined by the integral

$$\Gamma(n) = \int_0^\infty x^{n-1} e^{-x} dx, \quad n > 0 \tag{3.32}$$

which, of course, means

$$\Gamma(n+1) = \int_0^\infty x^n e^{-x} dx. \tag{3.33}$$

If you integrate (3.33) by-parts, then it is easy to show (you should do this) that

$$\Gamma(n+1) = n\Gamma(n), n > 0, \tag{3.34}$$

a result called the *functional equation* of the gamma function. Since

$$\Gamma(1) = \int_0^\infty e^{-x} dx = (-e^{-x}) \Big|_0^\infty = 1,$$

it then immediately follows that

$$\Gamma(n+1) = n!, \tag{3.35}$$

for n any positive integer. Thus, from (3.31), (3.33), and (3.35), we have

$$\mathcal{L}\{t^n\} = \frac{n!}{s^{n+1}} = \frac{\Gamma(n+1)}{s^{n+1}}. \tag{3.36}$$

Notice that for $n = 0$ (3.36) becomes

$$\mathcal{L}\{t^0\} = \mathcal{L}\{u(t)\} = \frac{1}{s},$$

a result we've already calculated directly in (3.21). The gamma function is due to Euler (are you surprised? — you shouldn't be, as he is *everywhere* in mathematics!), and he introduced it as a way to generalize the factorial function $n!$, from where n is a positive *integer*[9] to n being any real number.[10]

---

[9] $n\,! = n(n-1)(n-2)\ldots(2)(1)$ when n is a positive integer. For $n = 0$, write $n\,! = (n)(n-1)!$ and set $n = 1$. Then, $1\,! = (1)0!$ and so $0! = \frac{1!}{1} = \frac{1}{1} = 1$. To be emphatic about this, $0\,! \neq 0$ (!!!).

[10] The history of this can be found in Philip J. Davis, "Leonhard Euler's Integral: a historical profile of the gamma function," *American Mathematical Monthly*, December 1959, pp. 849–869.

What if n is not a positive integer? What if, for example, $n = -\frac{1}{2}$? That is, what is

$$\mathcal{L}\left\{\frac{1}{\sqrt{t}}\right\} = \frac{\Gamma\left(\frac{1}{2}\right)}{\sqrt{s}}?$$

Setting $n = -\frac{1}{2}$ in (3.32) we have

$$\Gamma\left(\frac{1}{2}\right) = \int_0^\infty x^{-\frac{1}{2}} e^{-x} dx.$$

Then, change variable to $x = y^2$ and so $dx = 2ydy$. Thus, as $y = 0$ when $x = 0$ and $y = \infty$ when $x = \infty$, we have

$$\Gamma\left(\frac{1}{2}\right) = \int_0^\infty (y^2)^{-\frac{1}{2}} e^{-y^2} 2ydy = 2\int_0^\infty \frac{1}{y} e^{-y^2} ydy$$

or,

$$\Gamma\left(\frac{1}{2}\right) = 2\int_0^\infty e^{-y^2} dy. \tag{3.37}$$

The integral in (3.37) is famous in science and engineering, one that all electrical engineers should really know how to do, if only as a 'badge of culture.' A more immediate reason, however, is that we'll encounter this integral, in the form of what is called the *error function*, when we get to transients in transmission lines. There we will need a transform pair that we haven't yet encountered. But, before we do that, we need to look more closely at the integral in (3.37). Instead of simply sending you off to the library to look in some pure math book (the usual engineering textbook practice), let me show you a spectacularly beautiful evaluation (due to some genius in the past unknown to me). The arguments are clever, and every electrical engineer should go through them at least once.

If we rewrite (3.37) using the dummy variable of integration x instead of y, we change nothing, and so it is certainly true that

$$\Gamma\left(\frac{1}{2}\right) = 2\int_0^\infty e^{-x^2} dx. \tag{3.38}$$

Now, multiply (3.37) and (3.38) together. Yes, an unexpected thing to do, I admit, but certainly if we *think* of doing it we *can* do it to get

$$\Gamma^2\left(\frac{1}{2}\right) = 4\int_0^\infty e^{-x^2} dx \int_0^\infty e^{-y^2} dy = 4\int_0^\infty \int_0^\infty e^{-\left(x^2+y^2\right)} dxdy. \tag{3.39}$$

Well, we do seem to be going in the wrong direction, don't we, from a one-dimensional definite integral that we can't do to a double definite integral that appears to be even worse.

But it's *not* worse! As I'll show you next, multiplying two 'undoable' expressions together has, in fact, given us a 'doable' entity. Where else but in math does something like that happen?

If you stare at that double integral in (3.39) long enough, perhaps the $x^2 + y^2$ in the exponent will remind you of the *physical* interpretation of x and y being the coordinates of the point in the plane that is distance $r = \sqrt{x^2 + y^2}$ from the origin. So, what (3.39) is *physically* doing is integrating e^{-r^2} over the entire first quadrant of the plane (that's what $0 \leq x \leq \infty$ and $0 \leq y \leq \infty$ covers). But what about that $dxdy$ in the integrand? Well, that's the *differential area* in Cartesian coordinates. But since r is the radius vector in *polar* coordinates, we should be using the differential area in polar coordinates, which you'll recall from freshman calculus is $rdrd\theta$, where θ is the angle the radius vector makes with the x-axis.[11] To integrate over the entire first quadrant we should use $0 \leq r \leq \infty$ and $0 \leq \theta \leq \frac{\pi}{2}$. So, with all this in mind, we rewrite the two-dimensional integral of e^{-r^2} over the entire first quadrant as

$$\Gamma^2\left(\frac{1}{2}\right) = 4\int_0^{\pi/2}\int_0^{\infty} e^{-r^2} rdrd\theta = 4\int_0^{\pi/2}\left\{\int_0^{\infty} e^{-r^2} rdr\right\}d\theta. \tag{3.40}$$

The θ-integral is easy, as there is no θ-dependency at all in the integrand. So,

$$\Gamma^2\left(\frac{1}{2}\right) = 4\frac{\pi}{2}\int_0^{\infty} e^{-r^2} rdr = 2\pi\int_0^{\infty} e^{-r^2} rdr. \tag{3.41}$$

And the r-integral in (3.41) is almost as easy to do, giving

$$\Gamma^2\left(\frac{1}{2}\right) = 2\pi\left(-\frac{1}{2}e^{-r^2}\right)\Big|_0^{\infty} = 2\pi\left(\frac{1}{2}\right) = \pi \tag{3.42}$$

and so, just like that, we have

$$\Gamma\left(\frac{1}{2}\right) = \sqrt{\pi}. \tag{3.43}$$

Looking back at what started us on this line of analysis, we have

$$\mathcal{L}\left\{\frac{1}{\sqrt{t}}\right\} = \frac{\sqrt{\pi}}{\sqrt{s}}. \tag{3.44}$$

We'll see this transform pair again, when we get to transmission lines.

[11]The more general way to handle such coordinate conversions in multi-dimensional integrals is to use what mathematicians call *Jacobians*, and for that I *will* send you off to the math library.

We can use (3.44), together with our earlier result in (3.7) that says

$$\text{if } f(t) \leftrightarrow F(s)$$
$$\text{then } \int_0^t f(x)dx \leftrightarrow \frac{1}{s}F(s)$$

to find the Laplace transform of $\sqrt{t}$. That is, since (3.44) says

$$\frac{1}{\sqrt{\pi t}} \leftrightarrow \frac{1}{\sqrt{s}}$$

then

$$\int_0^t \frac{1}{\sqrt{\pi x}}dx \leftrightarrow \frac{1}{s\sqrt{s}};$$

doing the integral gives us the pair

$$\sqrt{t} \leftrightarrow \frac{\sqrt{\pi}}{2s\sqrt{s}}. \tag{3.45}$$

Alternatively, we could just as easily use (3.8) which says since

$$\frac{1}{\sqrt{\pi t}} \leftrightarrow \frac{1}{\sqrt{s}}$$

then

$$t\frac{1}{\sqrt{\pi t}} = \frac{\sqrt{t}}{\sqrt{\pi}} \leftrightarrow -\frac{d}{ds}\left(\frac{1}{\sqrt{s}}\right) = -\frac{d}{ds}\left(s^{-1/2}\right)$$

or

$$\frac{\sqrt{t}}{\sqrt{\pi}} \leftrightarrow -\left(-\frac{1}{2}s^{-\frac{3}{2}}\right) = \frac{1}{2s\sqrt{s}}$$

or

$$\sqrt{t} \leftrightarrow \frac{\sqrt{\pi}}{2s\sqrt{s}}$$

and so again we have (3.45).

Our result in (3.36), for the transforms of the powers of time, is very useful in establishing yet another transform pair that we'll need when we get to transmission lines. Unlike the pair I mentioned earlier, involving the error function (which I'll delay developing until the end of this chapter), I'll end this section with the determination of the time function *f*(*t*) that pairs with the transform

$$F(s) = \frac{1}{\sqrt{s^2 + a^2}}.$$

The key idea is to recall the generalized form of the binomial theorem, which wasn't rigorously proven until as recently as 1826 by the Norwegian mathematician Niels Henrik Abel (1802–1829):

$$(1+x)^m = 1 + mx + \frac{m(m-1)}{2!}x^2 + \frac{m(m-1)(m-2)}{3!}x^3 + \ldots,$$

which holds for arbitrary m and $|x| < 1$.

Writing

$$F(s) = \frac{1}{s\sqrt{1+\left(\frac{a}{s}\right)^2}} = \frac{1}{s}\left[1+\left(\frac{a}{s}\right)^2\right]^{-1/2},$$

then with $x = \left(\frac{a}{s}\right)^2$ and $m = -\frac{1}{2}$ we have

$$F(s) = \frac{1}{s}\left[1 - \frac{1}{2}\left(\frac{a}{s}\right)^2 + \frac{\left(-\frac{1}{2}\right)\left(-\frac{3}{2}\right)}{2!}\left(\frac{a}{s}\right)^4 + \frac{\left(-\frac{1}{2}\right)\left(-\frac{3}{2}\right)\left(-\frac{5}{2}\right)}{3!}\left(\frac{a}{s}\right)^6 + \ldots\right].$$

The general term in this series for $F(s)$ is, for =0, 1, 2, 3, . . .,

$$\frac{1}{s}\left[(-1)^k \frac{\frac{(1)(3)(5)\ldots(2k-1)}{2^k}}{k!}\left(\frac{a}{s}\right)^{2k}\right] = (-1)^k \frac{(1)(3)(5)\ldots(2k-1)}{2^k k!} a^{2k} \frac{1}{s^{2k+1}}.$$

From (3.36) we have

$$\frac{t^n}{n!} \leftrightarrow \frac{1}{s^{n+1}}$$

and so, if $n + 1 = 2k + 1$ ($n = 2k$), then

$$\frac{t^{2k}}{(2k)!} \leftrightarrow \frac{1}{s^{2k+1}}.$$

Thus, the time function that pairs with the general term in $F(s)$ is

$$\begin{aligned}
&(-1)^k \frac{(1)(3)(5)\ldots(2k-1)}{2^k k!} a^{2k} \frac{t^{2k}}{(2k)!} \\
&= (-1)^k \frac{[(1)(3)(5)\ldots(2k-1)][(2)(4)\ldots(2k)]}{2^k k![(2)(4)\ldots(2k)]} \frac{(at)^{2k}}{(2k)!} \\
&= (-1)^k \frac{(2k)!}{2^k k!\left[2^k k!\right]} \frac{(at)^{2k}}{(2k)!} = (-1)^k \frac{(at)^{2k}}{2^{2k}(k!)^2} = (-1)^k \frac{\left(\frac{at}{2}\right)^{2k}}{(k!)^2}
\end{aligned}$$

and so

$$f(t) = \sum_{k=0}^{\infty} (-1)^k \frac{\left(\frac{at}{2}\right)^{2k}}{(k!)^2} \leftrightarrow \frac{1}{\sqrt{s^2 + a^2}}, t \geq 0. \tag{3.46}$$

Now, astonishingly, there is actually a mathematical name for the perhaps scary-looking (3.46). In advanced electrical engineering one soon encounters *Bessel's differential equation*[12]

$$x^2 \frac{d^2y}{dx^2} + x\frac{dy}{dx} + (x^2 - n^2)y = 0,$$

the solutions of which are called *Bessel functions* (there is a different one for every value of n). Their importance in applied mathematics is right behind that of the trigonometric and exponential functions.

In particular, a *Bessel function of the first kind of order n* has the power series expansion

$$J_n(x) = \sum_{k=0}^{\infty} (-1)^k \frac{\left(\frac{x}{2}\right)^{n+2k}}{k!(n+k)!}$$

and so we see that the $f(t)$ in (3.46) is $J_0(at)$, the Bessel function of the first kind of order zero. In fact, our analysis has given us a bonus, in that if we imagine that a is imaginary (that is, $a = b\sqrt{-1}$), then

$$J_0\left(bt\sqrt{-1}\right) \leftrightarrow \frac{1}{\sqrt{s^2 - b^2}}.$$

Instead of writing $J_0\left(bt\sqrt{-1}\right)$, however, the practice is to write $I_0(at)$ and to call this the *modified* Bessel function of the first kind of order zero. (Notice that since $\left(\frac{at}{2}\right)$ is raised to only *even* powers in the series of (3.46), then $I_0(at)$ is still purely real even though a is imaginary.) So,

$$J_0(at) \leftrightarrow \frac{1}{\sqrt{s^2 + a^2}} \tag{3.47}$$

and

$$I_0(at) \leftrightarrow \frac{1}{\sqrt{s^2 - a^2}}. \tag{3.48}$$

Both $J_0(x)$ and $I_0(x)$ are tabulated functions in printed tables, as well as available as callable functions in modern scientific software (like *MATLAB*).

---

[12] Named after the German mathematical astronomer Friedrich Wilhelm Bessel (1784–1846).

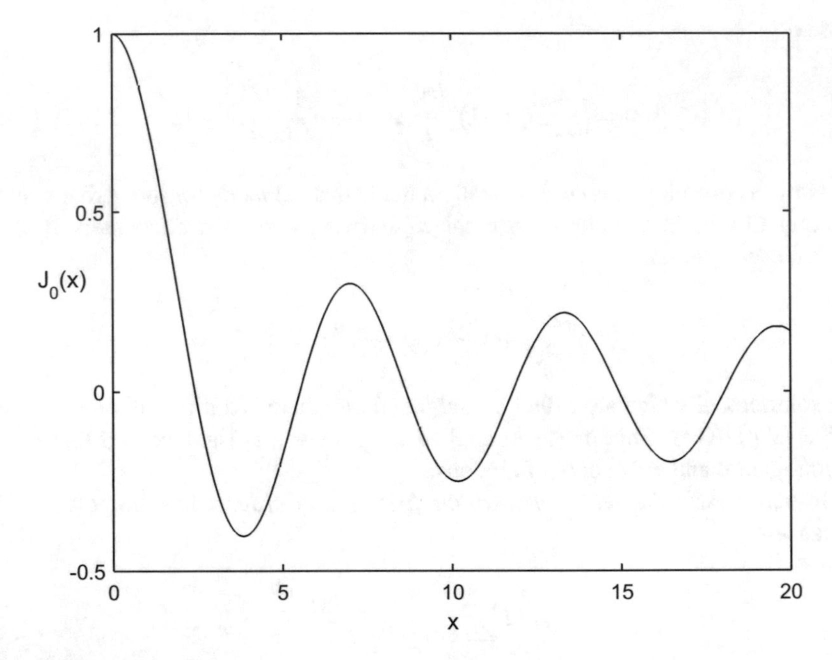

Fig. 3.6 Bessel function of the first kind of order zero

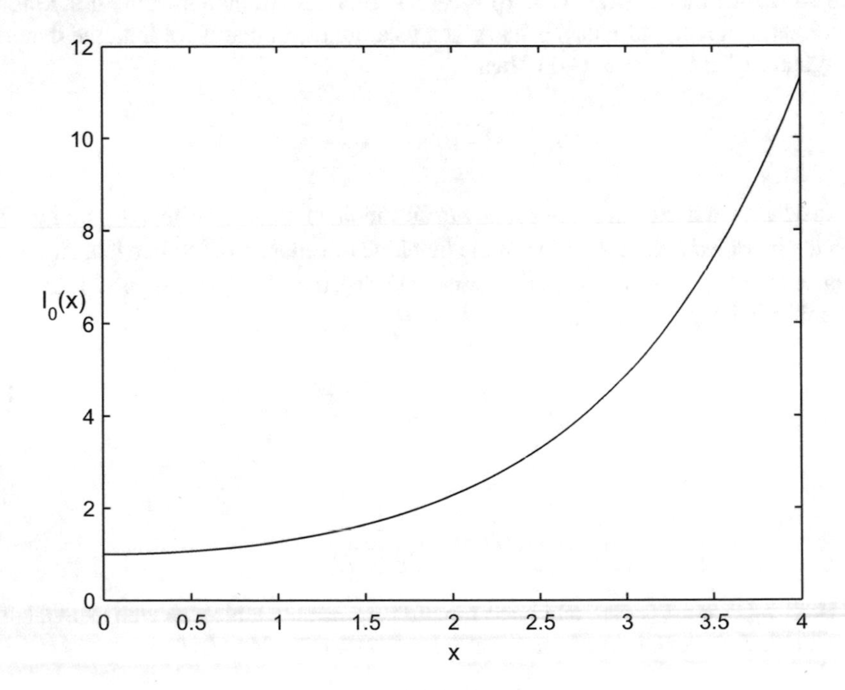

Fig. 3.7 Modified Bessel function of the first kind of order zero

Figures 3.6 and 3.7 show what $J_0(x)$ and $I_0(x)$ look like for $x > 0$, and they are clearly quite different. $J_0(x)$ is a decaying oscillation, while $I_0(x)$ is monotonic increasing.

3.5 Impulse Functions

In this section we encounter the strangest time function we'll deal with in this book, one that also (with no little irony) has the simplest Laplace transform: this is the *impulse* (or *Dirac delta*) function (see note 18 in the Preface). An impulse represents something that occurs 'all at once,' like a lightning strike near a transmission power line. To approximate such a thing, we can think of an impulse as the limiting case of a sequence of ever-narrower (in time) pulses of ever-increasing amplitude until, in the end, we have a pulse of zero duration and infinite amplitude. Think, for example, of the pulse shown in Fig. 3.8, which shows a narrow pulse of duration α and amplitude $1/\alpha$. The area bounded by this pulse is always 1, even as $\alpha \to 0$, and this observation will have important physical implications; for now, just tuck that fact away for later reference. To make explicit the role of α, the pulse is named $f_\alpha(t)$, and it is a perfectly ordinary, well-behaved function.

Now, imagine that we multiply $f_\alpha(t)$ with some *continuous* function $\phi(t)$, and then integrate the product over all t. That is, let's define the integral

$$I = \int_{-\infty}^{\infty} f_\alpha(t)\phi(t)dt = \int_0^\alpha \frac{1}{\alpha}\phi(t)dt \tag{3.49}$$

Then, suppose we let $\alpha \to 0$. In particular, the interval of integration (the duration of $f_\alpha(t)$) becomes arbitrarily small and, since $\phi(t)$ is *continuous*, then physically $\phi(t)$ cannot change by very much from the start of the integration interval to the end of that interval. Indeed, as $\alpha \to 0$ we can essentially treat $\phi(t)$ as constant over the entire interval, equal to $\phi(0)$, and so be able to pull it outside of the integral. So, in the limit $\alpha \to 0$ we see (3.49) becoming

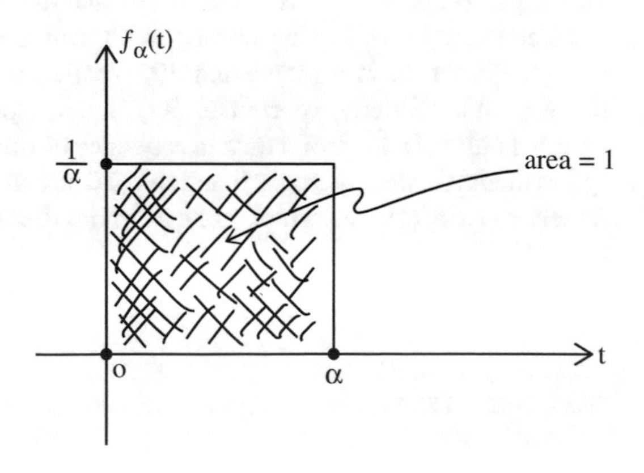

Fig. 3.8 As $\alpha \to 0$, the $f_\alpha(t)$ pulse becomes the *unit impulse*

$$\lim_{\alpha\to 0} I = \lim_{\alpha\to 0}\frac{1}{\alpha}\phi(0)\int_0^{\alpha} dt = \phi(0), \tag{3.50}$$

and the pulse in Fig. 3.8 becomes an *infinitely high spike* at $t = 0+$ that bounds unit area. Writing the impulse as

$$\delta(t) = \lim_{\alpha\to 0} f_\alpha(t),$$

we have what is called the sampling property of the impulse:

$$\int_{-\infty}^{\infty} \delta(t)\phi(t)dt = \phi(0). \tag{3.51}$$

From (3.51) we immediately have the Laplace transform of $\delta(t)$:

$$\mathcal{L}\{\delta(t)\} = \int_0^{\infty} \delta(t)e^{-st}dt = 1 \tag{3.52}$$

because $\phi(t) = e^{-st}$ and so $\phi(0) = 1$. This result is consistent with thinking of $\delta(t)$ as the derivative of the step $u(t)$, as the transform of $\delta(t)$ (that is, 1) is the transform of the step (that is, $\frac{1}{s}$) multiplied by s (an operation associated with time differentiation). This makes some intuitive sense, too, as $u(t)$ is a constant everywhere except at time $t = 0$, where it makes a jump of one in zero time (and now recall the definition of the derivative) — *but see Problem 3.13 for more on this subtle point*. This imagery, relating the step and the impulse was almost certainly suggested to Dirac when he first encountered the step function from the books on electrical circuits and electromagnetic wave theory by Oliver Heaviside. Equivalently, the step is the integral of the impulse:

$$u(t) = \int_{-\infty}^{t} \delta(x)dx = \begin{matrix} 1, & t > 0 \\ 0, & t < 0 \end{matrix} \tag{3.53}$$

which, you'll notice, avoids the issue of what the impulse is *at* $t = 0$.

Dirac knew he was being non-rigorous with arguments like the ones I've given here. As he wrote in a pioneering 1927 paper, published when he was still only 25 years old, "Strictly, of course, $\delta(x)$ is not a proper function of x, but can be regarded only as a limit of a certain sequence of functions. All the same one can use $\delta(x)$ as though it were a proper function . . . One can also use the [derivatives] of $\delta(x)$, namely $\delta'(x)$, $\delta''(x)$, . . ., which are ever more discontinuous and less 'proper' than $\delta(x)$ itself."[13]

[13]Paul Dirac, "The Physical Interpretation of Quantum Mechanics," *Proceedings of the Royal Society of London* (A113), January 1, 1927, pp. 621–641.

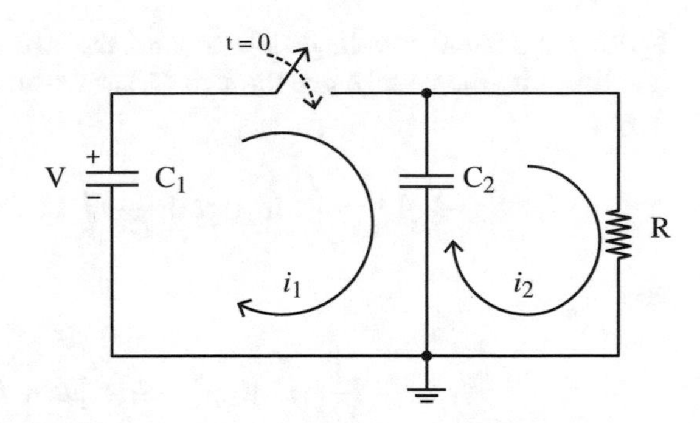

Fig. 3.9 What is $i_1(t)$ for $t \geq 0$?

One mathematician who almost certainly had Dirac's impulses and (3.53) in mind was Edward McShane (1904–1989) who, in his Presidential Address at the 1963 annual meeting of the American Mathematical Society, stated "There are in this world optimists who feel that any symbol that starts off with an integral sign must necessarily denote something that will have every property that they would like an integral to possess. This of course is quite annoying to us rigorous mathematicians: what is even more annoying is that by doing so they often come up with the right answer."[14] Like Dirac, McShane (who was a professor of mathematics at the University of Virginia) had an undergraduate degree in engineering, a background that allowed him to be willing to get his feet out of the 'cement of unrelenting rigor' before it completely hardened.[15] None of that, of course, is to shove sleazy mathematics under the rug, as the mathematics of impulses *has* been placed on a firm foundation since Dirac's use of them, through the work of the Russian mathematician Sergei Sobolev (1908–1988) and the French mathematician Laurent Schwartz (see note 19 in the Preface). We will here after use impulses without apology.

One of the concerns early analysts had with impulses is that they have infinite energy (see Problem 3.6) and, as you'll recall from Sect. 1.3, infinities in electrical circuits are generally considered to be red flags signaling that 'something is amiss.' Despite that, however, impulses do offer us explanations for what would otherwise be perplexing questions. Consider, for example, the circuit of Fig. 3.9 which shows a charged capacitor C_1 suddenly switched into a parallel connection with the *uncharged* C_2. The switch closing at $t = 0$ puts C_1 (with initial voltage drop V) in conflict with the initial zero voltage drop of C_2. This is a problem because we have argued that capacitor voltage drops cannot change instantly and so the two capacitors are clearly in conflict. The Laplace transform *automatically* shows us what happens (in particular, what the current $i_1(t)$ is in the initially charged capacitor) if we accept impulses.

---

[14]McShane's complete address, "Integrals Devised for Special Purposes," is in the *Bulletin of the American Mathematical Society*, September 1963, pp. 597–627.

[15]At the risk of enraging my more theoretical friends, let me quote (from an unknown source) this provocative thought: "Without engineering, physics and math are just philosophy."

Writing Kirchhoff's voltage law around the two loops[16] in the circuit, we have (starting with the voltage rise through C_1) as we go around each loop in a clockwise way

$$-V+\frac{1}{C_1}\int_0^t i_1(x)dx+\frac{1}{C_2}\int_0^t [i_1(x)-i_2(x)]dx=0 \tag{3.54}$$

and

$$\frac{1}{C_2}\int_0^t [i_2(x)-i_1(x)]dx+i_2R=0. \tag{3.55}$$

Laplace transforming (3.54) and (3.55), using our results in (3.7) and (3.11) and remembering that our mathematics is specifically restricted to $t>0$,

$$-\frac{V}{s}+\frac{I_1(s)}{C_1 s}+\frac{I_1(s)-I_2(s)}{C_2 s}=0 \tag{3.56}$$

and

$$\frac{I_2(s)-I_1(s)}{C_2 s}+I_2(s)R=0. \tag{3.57}$$

With just a little bit of easy algebra, (3.56) and (3.57) can be solved for $I_1(s)$:

$$I_1(s)=\left(\frac{C_1C_2}{C_1+C_2}\right)\frac{s+\frac{1}{RC_2}}{s+\frac{1}{R(C_1+C_2)}}. \tag{3.58}$$

Now, perform the long-division (which isn't really very long) of the ratio of polynomials of s in (3.58) to get

$$I_1(s)=\left(\frac{C_1C_2}{C_1+C_2}\right)\left[\frac{1+\frac{C_1}{RC_2(C_1+C_2)}}{s+\frac{1}{R(C_1+C_2)}}\right]. \tag{3.59}$$

We can now immediately write-down what $i_1(t)$ is, using our results in (3.12) and (3.52):

$$i_1(t)=\frac{C_1C_2}{C_1+C_2}V\left[\delta(t)+\frac{C_1}{RC_2(C_1+C_2)}e^{-\frac{t}{\{R(C_1+C_2)\}}}u(t)\right].$$

When the switch is closed in the circuit of Fig. 3.9 the voltage drops across C_1 and C_2 *do* become equal, *instantly*, because there is an *impulsive* current in C_1 that

[16]The use of loop currents in circuit analysis is due to the great Scottish mathematical physicist James Clerk Maxwell (1831–1879), who introduced the technique in his 1873 *Treatise on Electricity and Magnetism.*

transfers charge in the amount $\frac{C_1C_2}{C_1+C_2}V$ *instantly* into C_2, followed by an exponentially decaying current.

3.6 The Problem of the Reversing Current

In this section I'll show you how to solve the problem I teased you with in the Preface, concerning the circuit of Figure F1 (which I've reproduced, with some additional notation, in Fig. 3.10). I've assumed the battery voltage is 1 volt. I claimed in the Preface that after the switch is closed the current in the horizontal resistor first flows from right-to-left, declines to zero, and then reverses direction to flow left-to-right. We can show that much without writing even a single line of mathematics, but to find *when* the current reverses direction *will* take some math.

Here's why the current reverses. Both capacitors are initially uncharged, and so we have zero voltage drop across each C at $t = 0-$. Since the voltage drop across a capacitor cannot change instantly (unless the charging current is impulsive), then the voltage drop across each C must still be zero at $t = 0+$. Clearly, $x(0+) = u(0+) = 1$, and so $b(0-) = 0$ instantly jumps to $b(0+) = 1$ to keep the *drop* across the upper-vertical C at zero volts.[17] Since the lower-vertical C has its bottom-end directly connected to ground (0 volts), then its upper-end must also be at zero volts at $t = 0+$ to keep the *drop* at zero volts. Thus, $a(0+) = 0$. So, at $t = 0+$ the voltage drop across the horizontal resistor is

$$a(0+) - b(0+) = 0 - 1 = -1 \text{ volt} \tag{3.60}$$

and the current in that resistor flows right-to-left.

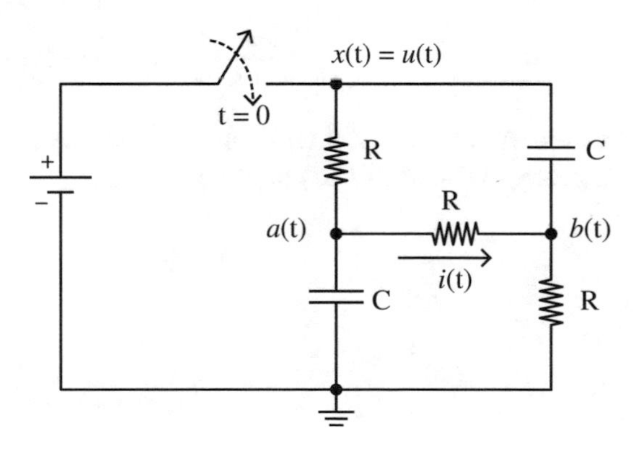

Fig. 3.10 When is $i(t) = 0$?

[17]The voltage at *each* end of a capacitor *can* change instantly, as long as *both* ends change by the same amount, thereby keeping the difference (the *drop*) unchanged.

After a long time, when the circuit has passed through its transient behavior and entered its d-c steady-state (and so no voltages or currents are changing), the only current path is through the series connection of the three R's, and therefore $a(\infty) = \frac{2}{3}$ and $b(\infty) = \frac{1}{3}$. Thus, at $t = \infty$, the voltage drop across the horizontal R is

$$a(\infty) - b(\infty) = \frac{2}{3} - \frac{1}{3} = \frac{1}{3}\,\text{volt} \tag{3.61}$$

and the current in that resistor flows left-to-right. The resistor current has reversed direction. Assuming the resistor current is a *continuous* function of time, then at some time after $t = 0+$ the current must equal zero. Here's how to calculate that time.

The plan of attack is straightforward: we'll simply calculate the voltage drop $a(t) - b(t)$ across the horizontal R and set it equal to zero. To that end, the Kirchhoff current law equations at the $a(t)$ and $b(t)$ nodes are

$$\frac{u(t) - a(t)}{R} = \frac{a(t) - b(t)}{R} + C\frac{da}{dt} \tag{3.62}$$

and

$$\frac{a(t) - b(t)}{R} + C\frac{d}{dt}\{u(t) - b(t)\} = \frac{b(t)}{R}. \tag{3.63}$$

Laplace transforming (3.62) and (3.63) gives

$$\frac{\frac{1}{s} - A}{R} = \frac{A - B}{R} + CsA \tag{3.64}$$

and

$$\frac{A - B}{R} + Cs\left(\frac{1}{s} - B\right) = \frac{B}{R}, \tag{3.65}$$

where $A = A(s) = \mathcal{L}\{a(t)\}$ and $B = B(s) = \mathcal{L}\{b(t)\}$. With just a bit of simple algebra, (3.64) and (3.65) become

$$A(RCs + 2) - B = \frac{1}{s} \tag{3.66}$$

and

$$A - B(RCs + 2) = -RC \tag{3.67}$$

where (3.66) and (3.67) have been written in the standard form for solution by Cramer's rule (determinants).

The system determinant of (3.66) and (3.67) is

$$D = \begin{vmatrix} RCs + 2 & -1 \\ 1 & -(RCs + 2) \end{vmatrix} = -(RCs + 2)^2 + 1$$

which, with just a touch of algebra, reduces to

$$D = -(RC)^2\left(s+\frac{3}{RC}\right)\left(s+\frac{1}{RC}\right). \tag{3.68}$$

Therefore,

$$A = \frac{\begin{vmatrix} \frac{1}{s} & -1 \\ -RC & -(RCs+2) \end{vmatrix}}{D} = -\frac{2RC+\frac{2}{s}}{D}$$

and

$$B = \frac{\begin{vmatrix} RCs+2 & \frac{1}{s} \\ 1 & -RC \end{vmatrix}}{D} = -\frac{RC(RCs+2)+\frac{1}{s}}{D}$$

and so

$$A-B = \frac{\left(-2RC-\frac{2}{s}\right)+RC(RCs+2)+\frac{1}{s}}{D}$$

which, again with a bit of algebra, becomes

$$A-B = \frac{1}{(RC)^2}\frac{1}{s\left(s+\frac{3}{RC}\right)\left(s+\frac{1}{RC}\right)} - \frac{s}{\left(s+\frac{3}{RC}\right)\left(s+\frac{1}{RC}\right)}. \tag{3.69}$$

Notice that (3.69) gives the correct values for $a(t) - b(t)$ when the initial and final value theorems are invoked:

$$\begin{aligned} a(0+)-b(0+) &= \lim_{s\to\infty} s(A-B) \\ &= \frac{1}{(RC)^2}\lim_{s\to\infty}\frac{1}{\left(s+\frac{3}{RC}\right)\left(s+\frac{1}{RC}\right)} \\ &\quad - \lim_{s\to\infty}\frac{s^2}{\left(s+\frac{3}{RC}\right)\left(s+\frac{1}{RC}\right)} = 0-1=-1, \end{aligned}$$

in agreement with (3.60), and

$$\begin{aligned} a(\infty)-b(\infty) &= \lim_{s\to 0} s(A-B) \\ &= \frac{1}{(RC)^2}\lim_{s\to 0}\frac{1}{\left(s+\frac{3}{RC}\right)\left(s+\frac{1}{RC}\right)} \\ &\quad - \lim_{s\to 0}\frac{s^2}{\left(s+\frac{3}{RC}\right)\left(s+\frac{1}{RC}\right)} = \frac{1}{(RC)^2}\frac{(RC)^2}{3} - 0 = \frac{1}{3}, \end{aligned}$$

in agreement with (3.61).

To return to the time domain, we need to make partial fraction expansions of

$$\frac{s}{\left(s+\frac{3}{RC}\right)\left(s+\frac{1}{RC}\right)}=\frac{k_1}{\left(s+\frac{3}{RC}\right)}+\frac{k_2}{\left(s+\frac{1}{RC}\right)} \tag{3.70}$$

and

$$\frac{1}{s\left(s+\frac{3}{RC}\right)\left(s+\frac{1}{RC}\right)}=\frac{k_3}{s}+\frac{k_4}{\left(s+\frac{3}{RC}\right)}+\frac{k_5}{\left(s+\frac{1}{RC}\right)}. \tag{3.71}$$

We'll use the same method on both expansions to get the k's. So, for (3.70), notice that multiplying through by $\left(s+\frac{3}{RC}\right)$ and then letting $s=-\frac{3}{RC}$ gives

$$\frac{s}{\left(s+\frac{1}{RC}\right)}\Big|_{s=-\frac{3}{RC}}=k_1+k_2\frac{s+\frac{3}{RC}}{s+\frac{1}{RC}}\Big|_{s=-\frac{3}{RC}}=k_1=\left(-\frac{3}{RC}\right)\left(-\frac{RC}{2}\right)=\frac{3}{2}.$$

Also, multiplying through by $\left(s+\frac{1}{RC}\right)$ and then letting $s=-\frac{1}{RC}$ gives

$$\frac{s}{\left(s+\frac{3}{RC}\right)}\Big|_{s=-\frac{1}{RC}}=k_1\frac{s+\frac{1}{RC}}{s+\frac{3}{RC}}\Big|_{s=-\frac{1}{RC}}+k_2=k_2=\left(-\frac{1}{RC}\right)\left(\frac{RC}{2}\right)=-\frac{1}{2}.$$

Repeating this procedure on (3.71), except now letting $s=0$, $s=-\frac{3}{RC}$, and $s=-\frac{1}{RC}$ we get

$$k_3=\frac{1}{\left(s+\frac{3}{RC}\right)\left(s+\frac{1}{RC}\right)}\Big|_{s=0}=\frac{(RC)^2}{3},$$

$$k_4=\frac{1}{s\left(s+\frac{1}{RC}\right)}\Big|_{s=-\frac{3}{RC}}=\frac{1}{\left(-\frac{3}{RC}\right)\left(-\frac{2}{RC}\right)}=\frac{(RC)^2}{6},$$

$$k_5=\frac{1}{s\left(s+\frac{3}{RC}\right)}\Big|_{s=-\frac{1}{RC}}=\frac{1}{\left(-\frac{1}{RC}\right)\left(\frac{2}{RC}\right)}=-\frac{(RC)^2}{2}.$$

Putting these k's into (3.70) and (3.71), and the results into (3.69), we arrive at

$$A-B=\frac{1}{(RC)^2}\left[\frac{\frac{(RC)^2}{3}}{s}+\frac{\frac{(RC)^2}{6}}{s+\frac{3}{RC}}-\frac{\frac{(RC)^2}{2}}{s+\frac{1}{RC}}\right]-\frac{\frac{3}{2}}{s+\frac{3}{RC}}+\frac{\frac{1}{2}}{s+\frac{1}{RC}}$$

$$=\frac{\frac{1}{3}}{s}+\frac{\frac{1}{6}}{s+\frac{3}{RC}}-\frac{\frac{1}{2}}{s+\frac{1}{RC}}-\frac{\frac{3}{2}}{s+\frac{3}{RC}}+\frac{\frac{1}{2}}{s+\frac{1}{RC}}$$

or,

$$A-B=\frac{\frac{1}{3}}{s}-\frac{\frac{8}{6}}{s+\frac{3}{RC}}\rightarrow\left[\frac{1}{3}-\frac{4}{3}e^{-\frac{3}{RC}t}\right]u(t). \tag{3.72}$$

Notice that (3.72) says

$$a(0+) - b(0+) = \frac{1}{3} - \frac{4}{3} = -1$$

and

$$a(\infty) - b(\infty) = \frac{1}{3} - 0 = \frac{1}{3}$$

in agreement with (3.60) and (3.61).

Now, to answer the question of *when* the current in the horizontal resistor equals zero, simply call the answer $t = T$ where

$$\frac{1}{3} - \frac{4}{3}e^{-\frac{3}{RC}T} = 0$$

and solve for T: that calculation is easily done to give

$$T = \frac{1}{3}RCln(4) = \frac{1}{3}RCln\left(2^2\right)$$

or, at last,

$$T = RC\frac{2}{3}\ \ln{(2)},$$

the answer given in note 6 of the Preface.

3.7 An Example of the Power of the Modern Electronic Computer

In this section of the chapter we'll work through the circuit in Fig. 3.11, using the Laplace transform. There are no initial currents or charges in the circuit and then, at time $t = 0$, a voltage pulse $v(t)$ with duration τ occurs with amplitude a (modeled by

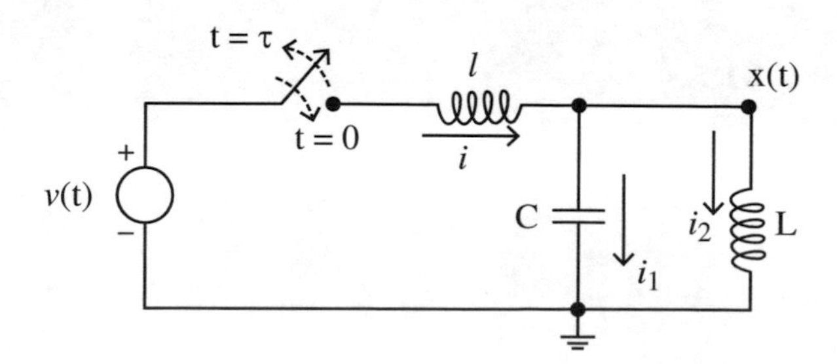

Fig. 3.11 What is $x(t)$ for a pulsed $v(t)$?

the switch closing at $t = 0$ and then opening at $t = \tau > 0$). Our problem is to determine the voltage $x(t)$ that appears across the parallel L, C combination.

From the given conditions, we know that $i(0+) = i_1(0+) = i_2(0+) = 0$, and that $x(0+) = 0$. The time-domain equations describing this circuit are:

$$v - x = l\frac{di}{dt},$$
$$i = i_1 + i_2,$$
$$i_1 = C\frac{dx}{dt},$$
$$x = L\frac{di_2}{dt}.$$

Transforming, these equations in time become the following in s:

$$V - X = slI,$$
$$I = I_1 + I_2,$$
$$I_1 = sCX,$$
$$X = sLI_2.$$

Next, some simple algebra:

$$I_2 = \frac{X}{sL}$$

and so

$$I = sCX + \frac{X}{sL} = X\left(sC + \frac{1}{sL}\right)$$

and therefore

$$V - X = slX\left(sC + \frac{1}{sL}\right)$$

or,

$$V = X + slX\left(sC + \frac{1}{sL}\right) = X\left[1 + sl\left(sC + \frac{1}{sL}\right)\right].$$

Thus,

$$X(s) = \frac{V(s)}{1 + s^2lC + \frac{l}{L}} - \frac{V(s)}{lC\left(s^2 + \frac{l+L}{LlC}\right)}. \tag{3.73}$$

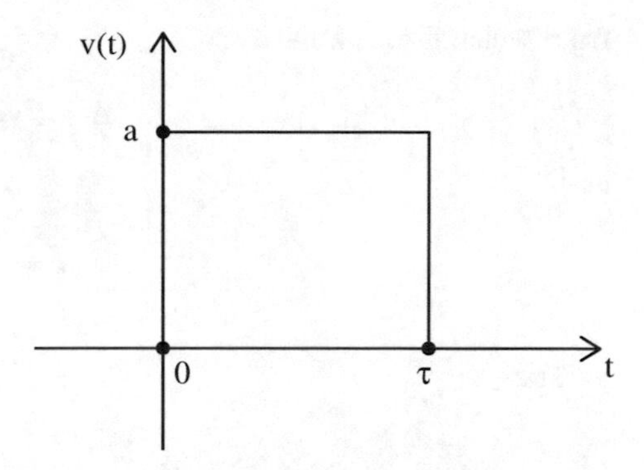

Fig. 3.12 The voltage pulse $v(t)$

The next obvious step is to determine $V(s)$. Figure 3.12 shows $v(t)$, a pulse of duration τ and amplitude a. We can write $v(t)$ as the difference between two step functions as follows:

$$v(t) = a[u(t) - u(t - \tau)]$$

and so, using (3.11) and (3.14),

$$V(s) = a\left[\frac{1}{s} - \frac{e^{-\tau s}}{s}\right]. \tag{3.74}$$

Putting (3.74) into (3.73) gives

$$X(s) = \frac{a\left[\frac{1}{s} - \frac{e^{-\tau s}}{s}\right]}{lC\left(s^2 + \frac{l+L}{LlC}\right)}$$

or, if we define

$$\omega_0^2 = \frac{l+L}{LlC}, \tag{3.75}$$

we have

$$X(s) = \frac{a}{lC}\left[\frac{1}{s\left(s^2 + \omega_0^2\right)} - \frac{e^{-\tau s}}{s\left(s^2 + \omega_0^2\right)}\right]. \tag{3.76}$$

If we can determine the time function that pairs with the first term in the square brackets, then the time function that goes with the second term is simply the first time function shifted (delayed) by τ. The first term is the result of the start of the pulse, and the second term is (of course) caused by the end of the pulse.

To determine these time functions, we'll make a partial fraction expansion: if A and B are some constants (to be determined next), then

$$\frac{1}{s\left(s^2 + \omega_0^2\right)} = \frac{A}{s} + \frac{Bs}{s^2 + \omega_0^2} = \frac{As^2 + A\omega_0^2 + Bs^2}{s\left(s^2 + \omega_0^2\right)}$$

from which it follows that

$$A + B = 0$$

and

$$A\omega_0^2 = 1.$$

Thus,

$$A = \frac{1}{\omega_0^2}, B = -A = -\frac{1}{\omega_0^2}$$

and so

$$\frac{1}{s\left(s^2 + \omega_0^2\right)} = \frac{\frac{1}{\omega_0^2}}{s} - \frac{\frac{1}{\omega_0^2}s}{s^2 + \omega_0^2}.$$

The time functions that pair with these two transforms are, from (3.11) and (3.15),

$$\frac{1}{\omega_0^2}u(t) - \frac{1}{\omega_0^2}\cos(\omega_0 t)u(t) = \frac{1}{\omega_0^2}[1 - \cos(\omega_0 t)]u(t)$$

and so we have our answer:

$$x(t) = \frac{a}{lC}\left\{\frac{1}{\omega_0^2}[1 - \cos(\omega_0 t)]u(t) - \frac{1}{\omega_0^2}[1 - \cos\{\omega_0(t - \tau)\}]u(t - \tau)\right\}$$

or, as

$$\frac{a}{lC\omega_0^2} = \frac{a}{lC\frac{l+L}{LlC}} = \frac{aL}{L+l},$$

$$\frac{x(t)}{a} = \frac{L}{L+l}\{[1 - \cos(\omega_0 t)]u(t) - [1 - \cos\{\omega_0(t - \tau)\}]u(t - \tau)\}. \qquad (3.77)$$

Before plotting (3.77), let's put in some numbers: $L = 5\ mH = 5{,}000\ \mu H$, $l = 40\ mH = 40{,}000\ \mu H$, and $C = 0.01\ \mu F$. Expressing the component units this way gives, you'll recall, time in units of microseconds. Finally,

$$\omega_0^2 = \frac{l+L}{LlC} = \frac{40,000 + 5,000}{5 \times 10^3 \times 4 \times 10^4 \times 10^{-2}} = 2.25 \times 10^{-2}$$

and so $\omega_0 = 0.15$. To understand what this physically means, ω_0 is the frequency at which the circuit will oscillate when the voltage pulse hits: if T is the period of the oscillation,

$$\omega_0 = \frac{2\pi}{T}$$

and so

$$T = \frac{2\pi}{\omega_0} = \frac{2\pi}{0.15} = 41.9\ \mu sec\ (\text{about } 23.9\ kHz).$$

Now, let me explain the significance of the title of this section. We could just take all of these numbers, plug them into (3.77) and, after picking a value for τ, plot $x(t)$ versus t. Modern computer software allows us to do something rather remarkable instead, something analysts when I was a student could only fantasize about. Imagine that we *haven't* derived (3.77), but only have gotten as far as (3.76). That is, we have $X(s)$, but haven't yet gathered the will and the strength to push on and simplify it by a partial fraction expansion (the grubby part, generally, of a transient analysis). Instead, wouldn't it be neat if we could just type $X(s)$ into a computer and let *it* do all that awful arithmetic work, including generating the final plot? *Sure* it would, and today we *can*!

Different software packages will of course work differently, but the following code **pulse.m** shows one way to do it in *MATLAB*, using its powerful *ilaplace* command for performing inverse Laplace transforming. The result generated by **pulse.m** is Fig. 3.13 (for a $\tau = 43\ \mu sec$ voltage pulse, picked to be a bit larger than T but otherwise arbitrarily), showing a peak transformer voltage of about 22% of the pulse amplitude. The details of the code are unimportant for us (this is *not* a book on

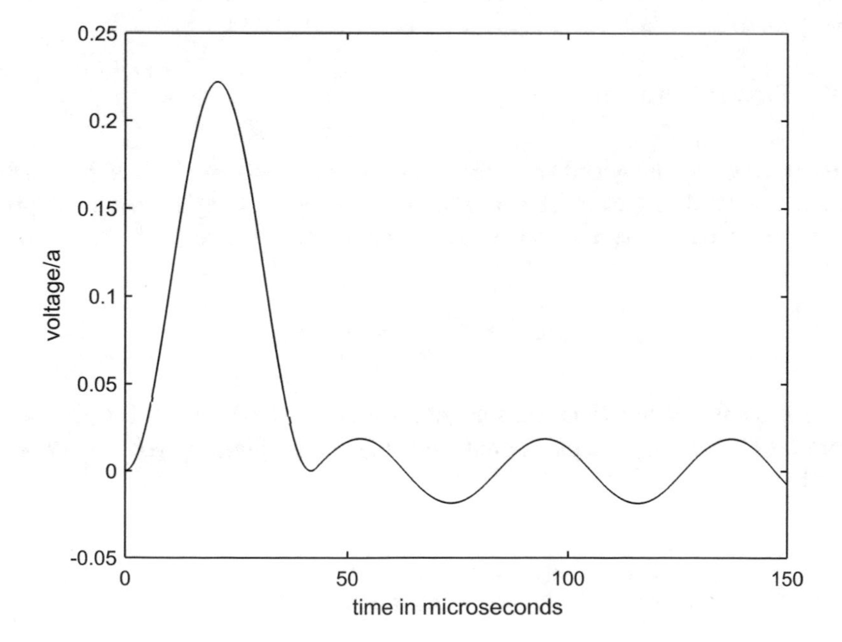

Fig. 3.13 The response of the circuit in Fig. 3.11

MATLAB programming!), and the only thing I want you to appreciate is how *brief* is the code, and its *generality*. Just change the circuit element values and the argument of the *ilaplace* command (and the time span in the *linspace* command) and the same code 'solves' just about any other transient problem involving a transform in the form of the ratio of two polynomials in s. In the next chapter I'll show you a few more examples of essentially the same code in action, with circuits more complicated than the one in Fig. 3.11, and how using a computer can save you from tidal waves of brain-crushing arithmetic.

```
%pulse.m/created by PJNahin for Electrical Transients (7/8/
2017)
L=5000;l=40000;C=0.01;
f=1/(l*C);
d=input('What is tau?');
w02=(l+L)/(L*l*C);
syms s t
h=ilaplace((1/(s*(s^2+w02)))*(1-exp(-d*s)));
h=h*f;
t=linspace(0,150,250);
y=subs(h);
v=vpa(y);
plot(t,v,'-k')
xlabel('time in microseconds')
ylabel('voltage/a')
```

3.8 Puzzle Solution

Have you been thinking hard about the second puzzle I described in Sect. 1.3? Just to remind you (and to generalize it a bit), here it is again. Imagine that we apply a voltage pulse $v(t)$ to the terminals of a one-ohm resistor, where

$$v(t) = \begin{matrix} 0, & t < 0 \\ c^{m-1}, & 0 < t < c \\ 0, & t > c \end{matrix}$$

where c is any positive finite constant and $0 < m < \frac{1}{2}$ (in Sect. 1.3 I used $m = \frac{1}{5}$). From Ohm's law we have the current $i(t)$ in the resistor given by $v(t) = i(t)R$ or, as $R = 1$,

$$i(t) = \begin{matrix} 0, & t < 0 \\ c^{m-1}, & 0 < t < c\,. \\ 0, & t > c \end{matrix}$$

That is, $i(t)$ is a finite-valued pulse that is non-zero only for a finite length of time. Now, the total energy dissipated as heat by the resistor due to this current pulse is

$$W = \int_{-\infty}^{\infty} i^2(t)dt = \int_0^c c^{2m-2}dt = c^{2m-2}c$$

or,

$$W = c^{2m-1}.$$

Also, the total electric charge that passes through the resistor is

$$Q = \int_{-\infty}^{\infty} i(t)dt = \int_0^c c^{m-1}dt = c^{m-1}c$$

or,

$$Q = c^m.$$

So, for m in the interval $0 < m < \frac{1}{2}$ we have

$$\lim_{c\to 0} W = \infty$$

and

$$\lim_{c\to 0} Q = 0.$$

These last two limits are quite strange. As $c \to 0$, the pulse-like current becomes ever briefer in duration but ever larger in amplitude. Since $Q \to 0$ as $c \to 0$ then, even though the current amplitude blows-up the pulse duration becomes shorter 'even faster' and so the total charge transported through the resistor goes to zero. And yet, the total dissipated *energy* blows-up. Indeed, since $W \to \infty$ as $c \to 0$ the resistor will instantly vaporize because all that infinite energy is delivered in zero time. But how can *that* be when, as $c \to 0$, there is no charge transported? Paradox!

The way to escape from this quandary is essentially the same as that for the two-capacitor/missing energy paradox we treated earlier in Chap. 1. There we argued that our original circuit of two capacitors was unrealistically simple. We addressed that issue by adding a vanishingly small resistance. To remove this new paradox, we again need to add something more to our minimal circuit of a single resistor; this time the addition is the self-inductance of the very wire that forms the circuit itself. That is, the equation describing our circuit is not $v = iR$ but rather, for some $L > 0$,

$$v = iR + L\frac{di}{dt}.$$

The reason for the additional term is that the current $i(t)$ generates a magnetic flux field that threads through the closed circuit loop, generating a so-called *back emf*[18] (*electromotive force*, or voltage), as described by Faraday's law of induction. As $c \to 0$ the current becomes ever more impulse-like, which means the magnetic flux is changing ever more rapidly, and the new self-induction term becomes non-negligible.

The Laplace transform of this modified equation is

$$V(s) = I(s) + L[sI(s) - i(0+)] = I(s) + LsI(s)$$

and so[19]

$$I(s) = \frac{V(s)}{1 + sL} = \frac{1}{L}\frac{V(s)}{s + \frac{1}{L}}.$$

Since

$$v(t) = c^{m-1}[u(t) - u(t - c)]$$

then

$$V(s) = \frac{c^{m-1}}{s}[1 - e^{-sc}]$$

and so

$$I(s) = \frac{c^{m-1}}{L}\frac{1 - e^{-sc}}{s\left(s + \frac{1}{L}\right)} = \frac{c^{m-1}}{L}\left[\frac{1}{s\left(s + \frac{1}{L}\right)} - \frac{e^{-sc}}{s\left(s + \frac{1}{L}\right)}\right].$$

The second term in the square brackets on the right is just a time-shifted version of the first term in the square brackets. So, for that first term,

$$\frac{1}{s\left(s + \frac{1}{L}\right)} = L\left[\frac{1}{s} - \frac{1}{s + \frac{1}{L}}\right]$$

[18]The term comes from *Lenz's law* – after the Estonian physicist Heinrich Lenz (1804–1865) — who observed that Faraday's induced voltage is always such as to *oppose* the changing flux that created the induced voltage. Lenz's law is actually a manifestation of the conservation of energy (what would be the consequence if the induced voltage *enhanced* the changing flux?).

[19]$i(0+)$, the current in the circuit just after the current starts, is zero because the current in the 'inductor' cannot change instantly and $i(0-)$, the current just before the voltage pulse starts, is zero.

or,

$$\frac{1}{s\left(s+\frac{1}{L}\right)} \leftrightarrow L\left[1-e^{-\frac{t}{L}}\right]u(t).$$

Thus,

$$\frac{c^{m-1}}{L}\frac{1}{s\left(s+\frac{1}{L}\right)} \leftrightarrow c^{m-1}\left[1-e^{-\frac{t}{L}}\right]u(t)$$

and so the resistor current is

$$i(t) = c^{m-1}\left[1-e^{-\frac{t}{L}}\right]u(t) - c^{m-1}\left[1-e^{-\frac{(t-c)}{L}}\right]u(t-c).$$

The voltage pulse is limited to a finite interval of time, but the resulting current exists for all $t > 0$.

Clearly, and by inspection, $\lim_{c \to 0} i(t) = 0$ and so $\lim_{c \to 0} Q = 0$. But what of W? We have *during the voltage pulse* (pay close attention to the step functions), when $0 < t < c$,

$$i(t) = c^{m-1}\left[1-e^{-\frac{t}{L}}\right]$$

and so the energy dissipated *during the voltage pulse* is

$$W_p = \int_0^c i^2(t)dt = c^{2m-2}\int_0^c \left[1-e^{-\frac{t}{L}}\right]^2 dt$$

which is, if you're careful doing the integral,

$$W_p = c^{2m-1} - 2Lc^{2m-2}\left(1-e^{-\frac{c}{L}}\right) + \frac{1}{2}Lc^{2m-2}\left(1-e^{-\frac{2c}{L}}\right).$$

The current continues to exist after the voltage pulse, too, of course, and is given by, when $t > c$,

$$i(t) = c^{m-1}\left[e^{-\frac{t-c}{L}} - e^{-\frac{t}{L}}\right]$$

and so the energy dissipated *after* the voltage pulse is

$$W_{ap} = \int_c^\infty i^2(t)dt = c^{2m-2}\int_c^\infty \left[e^{-\frac{t-c}{L}} - e^{-\frac{t}{L}}\right]^2 dt$$

which is

$$W_{ap} = c^{2m-2}L\left[\frac{1}{2} - e^{-\frac{c}{L}} + \frac{1}{2}e^{-\frac{2c}{L}}\right].$$

The total energy dissipated is therefore

$$W = W_p + W_{ap} = c^{2m-1} - Lc^{2m-2}\left(1 - e^{-\frac{c}{L}}\right) = c^{2m-1} - T,$$

where

$$T = Lc^{2m-2}\left(1 - e^{-\frac{c}{L}}\right).$$

Notice that $T = 0$ if $L = 0$ and so $W = c^{2m-1}$ if $L = 0$, which blows-up as $c \to 0$ for $0 < m < \frac{1}{2}$ and we again have our paradox. It is clear then that $T > 0$ for $L > 0$ will be the origin of our salvation (*if* there *is* salvation to be had).

To see what happens to T as $c \to 0$, expand the exponential in its infinite power series to get

$$\begin{aligned} T &= Lc^{2m-2}\left[1 - \left\{1 + \frac{\left(-\frac{c}{L}\right)}{1!} + \frac{\left(-\frac{c}{L}\right)^2}{2!} + \frac{\left(-\frac{c}{L}\right)^3}{3!} + \frac{\left(-\frac{c}{L}\right)^4}{4!} + \ldots\right\}\right] \\ &= Lc^{2m-2}\left[\frac{c}{L} - \frac{1}{2}\left(\frac{c}{L}\right)^2 + \frac{1}{6}\left(\frac{c}{L}\right)^3 - \frac{1}{24}\left(\frac{c}{L}\right)^4 + \ldots\right] \\ &= c^{2m-1} - Lc^{2m-2}\left[\frac{1}{2}\left(\frac{c}{L}\right)^2 - \frac{1}{6}\left(\frac{c}{L}\right)^3 + \frac{1}{24}\left(\frac{c}{L}\right)^4 - \ldots\right] \end{aligned}$$

and so

$$\begin{aligned} W &= Lc^{2m-2}\left[\frac{1}{2}\left(\frac{c}{L}\right)^2 - \frac{1}{6}\left(\frac{c}{L}\right)^3 + \frac{1}{24}\left(\frac{c}{L}\right)^4 - \ldots\right] \\ &= Lc^{2m}\left[\frac{1}{2}\frac{1}{L^2} - \frac{1}{6}\frac{c}{L^3} + \frac{1}{24}\frac{c^2}{L^4} - \ldots\right] \\ &= \frac{c^{2m}}{L}\left[\frac{1}{2} - \frac{1}{6}\frac{c}{L} + \frac{1}{24}\frac{c^2}{L^2} - \ldots\right] \end{aligned}$$

which clearly goes to zero as $c \to 0$ for any $L > 0$ and any m in the interval $0 < m < \frac{1}{2}$. (If this *isn't* 'clear,' get hold of any freshman calculus book and look-up the convergence behavior for an infinite series in powers of $\left(\frac{c}{L}\right)$ with alternating signs). Paradox removed!

3.9 The Error Function and the Convolution Theorem

To end this chapter, I'll next show you two additional results in Laplace transform theory. For our first result, you'll see how the integral of (3.37) helps us find a particularly important transform pair, one we'll find invaluable when we get to transients in transmission lines. Specifically, if we define

$$erf(t) = \frac{2}{\sqrt{\pi}} \int_0^t e^{-u^2} du \tag{3.78}$$

then we'll derive the pair

$$1 - erf\left(\frac{a}{2\sqrt{t}}\right) \leftrightarrow \frac{1}{s} e^{-a\sqrt{s}} \tag{3.79}$$

where *erf* is the *error function* I mentioned back in Sect. 3.4. It's clear that $erf(0) = 0$, and also that the $\frac{2}{\sqrt{\pi}}$ factor in front of the integral in (3.78) makes $erf(\infty) = 1$ (because of (3.38) and (3.43)). This range of values is attractive because the error function plays a very big role in probability theory (and, of course, all probabilities are from 0 to 1) and, while our transient analyses here have nothing to do with probability, that's why $erf(t)$ is what it is. Now, to derive (3.79), I'll start by computing the Laplace transform of

$$f(t) = \frac{e^{-a^2/4t}}{\sqrt{\pi t^3}} \tag{3.80}$$

which is, admittedly, probably not an obvious thing to do! So, how *do* I know to do this?

The answer is that once, long ago, I read the famous 1822 book *Analytical Theory of Heat* by the French mathematical physicist Joseph Fourier (1768–1830), the same man who gave electrical engineers their beloved Fourier series (and Fourier transforms, too, which are closely related to Laplace transforms). That masterpiece works out, *all in the time domain*, how heat flows in solids, with the fundamental physics behind it all called the *diffusion* (or *heat*) equation. In its simplest form, that equation is a partial differential equation in two variables (one-dimension of space, and one of time) and, by some wondrous good fortune, it is also the underlying physics of the first (1855) transient analysis made of a transmission line (a submarine telegraph cable). Heat and electricity, both (under the right circumstances) obey the same equation, and that curious time function in (3.80) appeared in Fourier's book *decades* before William Thomson (look back at Sect. 1.4) remembered *his* reading of Fourier's book and used it to guide his analysis of what eventually resulted in the famous Atlantic Cable Project of 1866, a topic we'll take-up in detail in Chap. 5. So, please understand that I have *not* just pulled (3.80) out of thin air.

In any case, what we have is

$$F(s) = \int_0^\infty f(t)e^{-st}\,dt = \int_0^\infty \frac{e^{-a^2/4t}}{\sqrt{\pi t^3}}e^{-st}\,dt. \tag{3.81}$$

Next, let's make the change of variable

$$\tau^2 = \frac{a^2}{4t}$$

and so

$$\tau = \frac{a}{2t^{1/2}}$$

which says

$$dt = -\sqrt{t^3}\frac{4}{a}\,d\tau.$$

With this, (3.81) becomes

$$F(s) = \int_\infty^0 \frac{e^{-\tau^2}}{\sqrt{\pi t^3}}e^{-s\frac{a^2}{4\tau^2}}\left(-\sqrt{t^3}\frac{4}{a}\,d\tau\right) = -\frac{4}{a\sqrt{\pi}}\int_\infty^0 e^{-\left[\tau^2+s\frac{a^2}{4\tau^2}\right]}d\tau$$

$$= \frac{4}{a\sqrt{\pi}}\int_0^\infty e^{-\left[\tau^2+s\frac{a^2}{4\tau^2}+a\sqrt{s}-a\sqrt{s}\right]}d\tau = \frac{4}{a\sqrt{\pi}}e^{-a\sqrt{s}}\int_0^\infty e^{-\left[\tau^2\ -a\sqrt{s}\ +\left(s\frac{a\sqrt{s}}{2\tau}\right)^2\right]}d\tau$$

$$= \frac{4}{a\sqrt{\pi}}e^{-a\sqrt{s}}\int_0^\infty e^{-\left[\tau-\frac{a\sqrt{s}}{2\tau}\right]^2}d\tau$$

or,

$$F(s) = \frac{4}{a\sqrt{\pi}}e^{-a\sqrt{s}}\int_0^\infty e^{-\left[\tau-\frac{b}{\tau}\right]^2}d\tau \tag{3.82}$$

where

$$b = \frac{a\sqrt{s}}{2}.$$

Now, concentrate on the integral, alone, in (3.82). Multiplying out the exponent of the integrand, we have

$$\int_0^\infty e^{-\left(\tau^2-2b+\frac{b^2}{\tau^2}\right)}d\tau = e^{2b}\int_0^\infty e^{-\left(\tau^2+\ \frac{b^2}{\tau^2}\right)}d\tau = e^{2b}I(b) \tag{3.83}$$

where

$$I(b) = \int_0^{\infty} e^{-\left(\tau^2 + \frac{b^2}{\tau^2}\right)} d\tau. \tag{3.84}$$

The integral $I(b)$, despite its perhaps complicated appearance, can be evaluated as follows.

Differentiating (3.84) with respect to b (using Leibniz's formula, as in note 3 of Chap. 1),

$$\frac{dI}{db} = \int_0^{\infty} \left(-\frac{2b}{\tau^2}\right) e^{-\left(\tau^2 + \frac{b^2}{\tau^2}\right)} d\tau.$$

Now, make the change of variable

$$\tau = \frac{b}{u}$$

which says

$$d\tau = -\frac{b}{u^2} du$$

and so

$$\frac{dI}{db} = \int_{\infty}^{0} \left(-\frac{2u^2}{b}\right) e^{-\left(\frac{b^2}{u^2} + u^2\right)} \left(-\frac{b}{u^2} du\right) = -2 \int_0^{\infty} e^{-\left(\frac{b^2}{u^2} + u^2\right)} du = -2I(b).$$

Thus,

$$\frac{dI}{I} = -2db$$

which immediately integrates to

$$I(b) = Ce^{-2b}$$

where C is some constant. In fact,

$$I(0) = C = \int_0^{\infty} e^{-\tau^2} d\tau = \frac{\sqrt{\pi}}{2}$$

because, from (3.38) and (3.43), we know

$$2\int_0^{\infty} e^{-x^2} dx = \Gamma\left(\frac{1}{2}\right) = \sqrt{\pi}.$$

So,

$$I(b) = \frac{\sqrt{\pi}}{2} e^{-2b}$$

and putting that into (3.83) — which is, in fact, the integral in (3.82) — we have

$$\int_0^\infty e^{-\left[\tau - \frac{b}{\tau}\right]^2} d\tau = \frac{\sqrt{\pi}}{2}$$

a result (perhaps surprisingly) *independent* of b.

In any case, and just like that, we have the $F(s)$ of (3.82) for the $f(t)$ in (3.80):

$$F(s) = \frac{4e^{-a\sqrt{s}}}{a\sqrt{\pi}} \left(\frac{\sqrt{\pi}}{2} \right) = \frac{2}{a} e^{-a\sqrt{s}}$$

which gives us the pair

$$\frac{e^{-a^2/4t}}{\sqrt{\pi t^3}} \leftrightarrow \frac{2}{a} e^{-a\sqrt{s}}$$

which is equivalent to

$$\frac{ae^{-a^2/4t}}{2\sqrt{\pi t^3}} \leftrightarrow e^{-a\sqrt{s}}. \tag{3.85}$$

Finally, recall (3.7) which says given the pair $f(t) \leftrightarrow F(s)$ then

$$\int_0^\infty f(x)dx = \frac{F(s)}{s}.$$

Thus,

$$\frac{a}{2\sqrt{\pi}} \int_0^\infty \frac{e^{-a^2/4x}}{\sqrt{x^3}} dx \leftrightarrow \frac{e^{-a\sqrt{s}}}{s}. \tag{3.86}$$

In the integral of (3.86) make the change of variable

$$u = \frac{a}{2\sqrt{x}}$$

which says

$$dx = -\frac{4\sqrt{x^3}}{a} du$$

and so

$$\frac{a}{2\sqrt{\pi}}\int_{\infty}^{\frac{a}{2\sqrt{t}}}\frac{e^{-u^2}}{\sqrt{x^3}}\left(-\frac{4\sqrt{x^3}}{a}du\right)=\frac{2}{\sqrt{\pi}}\int_{\frac{a}{2\sqrt{t}}}^{\infty}e^{-u^2}du$$
$$=\frac{2}{\sqrt{\pi}}\left[\int_0^{\infty}e^{-u^2}du-\int_0^{\frac{a}{2\sqrt{t}}}e^{-u^2}du\right]$$

or, recalling (3.78), we have

$$\frac{2}{\sqrt{\pi}}\left[\frac{\sqrt{\pi}}{2}-\frac{\sqrt{\pi}}{2}erf\left(\frac{a}{2\sqrt{t}}\right)\right]\leftrightarrow\frac{e^{-a\sqrt{s}}}{s}$$

and so, at last, we have the claimed pair of (3.79):

$$1-erf\left(\frac{a}{2\sqrt{t}}\right)\leftrightarrow\frac{1}{s}e^{-a\sqrt{s}}.$$

As a quick partial check, since $erf(0)=0$ then, if $a=0$, this pair reduces to

$$1=u(t)\leftrightarrow\frac{1}{s}$$

which is our earlier result of (3.11) for the unit step function.

For the second result of this section, suppose we have two time functions, $f(t)$ and $g(t)$, with Laplace transforms $F(s)$ and $G(s)$, respectively. What time function pairs with the product $F(s)G(s)$? The answer is what is called the *convolution* of $f(t)$ and $g(t)$, written as $f(t) * g(t)$. That is,

$$f(t)*g(t)=\int_0^t f(t-p)g(p)dp\leftrightarrow F(s)G(s). \tag{3.87}$$

To prove (3.87), called the *convolution theorem*, we'll simply directly calculate the Laplace transform of the convolution integral. Before we start, however, we'll make one extremely helpful modification. The integration range on p runs from 0 to t. If p should run beyond t, then $f(t-p)$ would have a negative argument, and you'll recall that we are interested only in the behavior of time functions with positive arguments. So, if we replace $f(t-p)$ with $f(t-p)u(t-p)$ we change nothing because $u(t-p)=0$ for $p>t$. But with that change we can now let p run from 0 to ∞. That is,

$$\int_0^t f(t-p)g(p)dp=\int_0^{\infty}f(t-p)u(t-p)g(p)dp$$

and what we are going to compute is

$$\mathcal{L}\left\{\int_0^\infty f(t-p)u(t-p)g(p)dp\right\} = \int_0^\infty \left\{\int_0^\infty f(t-p)u(t-p)g(p)dp\right\}e^{-st}dt.$$

This looks pretty scary, but much less so if we reverse the order of integration and write

$$\mathcal{L}\left\{\int_0^\infty f(t-p)u(t-p)g(p)dp\right\} = \int_0^\infty g(p)\left\{\int_0^\infty f(t-p)u(t-p)e^{-st}dt\right\}dp.$$

That is, everything with a t in it has to stay in the inner integral (which is with respect to t), while everything with *only* p in it can come out to the outer integral. For pure mathematicians, this step of reversing the order of integration demands justification, which means showing that the integrals involved are uniformly convergent. As engineers we are going to skip that step (all the while realizing we are making a *big* assumption) and instead carry-out what is called a *formal analysis* (a euphemism admitting we are not being pure).

Now, in the inner integral make the change of variable $t - p = z$ (and so $dt = dz$). Thus,

$$\mathcal{L}\left\{\int_0^\infty f(t-p)u(t-p)g(p)dp\right\} = \int_{-p}^\infty g(p)\left\{\int_{-p}^\infty f(z)u(z)e^{-s(p+z)}dz\right\}dp.$$

The inner integral in this last result can be written as

$$\int_{-p}^\infty f(z)u(z)e^{-s(p+z)}dz = e^{-sp}\int_0^\infty f(z)e^{-sz}dz = e^{-sp}F(s).$$

Note, carefully, that we can bring the lower integration limit up from $-p$ to 0 in the integral because of the $u(z)$ in the integrand. So,

$$\mathcal{L}\left\{\int_0^\infty f(t-p)u(t-p)g(p)dp\right\} = \int_0^\infty g(p)e^{-sp}F(s)dp = F(s)\int_0^\infty g(p)e^{-sp}dp$$

or, just like that,

$$\mathcal{L}\left\{\int_0^\infty f(t-p)g(p)dp\right\} = F(s)G(s)$$

and (3.87) is established.

As an example to illustrate the power of the convolution theorem, let's use it to find the time function that pairs with

$$\frac{1}{(s-1)\sqrt{s}} = \left(\frac{1}{s-1}\right)\left(\frac{1}{\sqrt{s}}\right) = F(s)G(s) \tag{3.88}$$

where

$$F(s) = \frac{1}{s-1}, \quad G(s) = \frac{1}{\sqrt{s}}.$$

We know from (3.12) that

$$f(t) = e^t u(t)$$

and from (3.44) that

$$g(t) = \frac{1}{\sqrt{\pi t}} u(t),$$

where I've explicitly included a step function $u(t)$ in each time function to remind us that time functions have meaning for us only for $t > 0$.

The convolution theorem tells us that the time function we are after is, from (3.87),

$$f(t) * g(t) = \int_0^t e^{(t-p)} u(t-p) \frac{1}{\sqrt{\pi p}} u(p) dp = \frac{e^t}{\sqrt{\pi}} \int_0^t \frac{e^{-p}}{\sqrt{p}} dp.$$

I've dropped the step functions in the right-most integral because the product $u(t-p)u(p) = 1$ as p varies from 0 to t in the integration interval, and is zero for $p > t$. Now, make the change of variable

$$v = \sqrt{p} \text{ (and so } dp = 2v dv)$$

to get

$$f(t) * g(t) = \frac{e^t}{\sqrt{\pi}} \int_0^{\sqrt{t}} \frac{e^{-v^2}}{v} 2v dv = e^t \frac{2}{\sqrt{\pi}} \int_0^{\sqrt{t}} e^{-v^2} dv$$

and so, from (3.78), we have our answer:

$$f(t) * g(t) = e^t \text{erf}\left(\sqrt{t}\right) \leftrightarrow \frac{1}{(s-1)\sqrt{s}}.$$

This long chapter has been pretty mathematical, but I hope the circuit examples in it have justified for you the worth of making the effort to study the Laplace transform. The mathematician Ruel V. Churchill (see note 10 in the Preface) opened his book on the operational calculus with these words: "Since the time of its introduction the operational calculus of Oliver Heaviside has held a prominent place in the treatment of problems in electric circuits . . . But in its original form this method rested on rules of procedure that had no satisfactory logical justification. Nor were the rules always reliable.[20] *The modern form of this operational calculus consists of the use of the Laplace transform* [my emphasis]." The rest of this book is devoted to illustrating what Professor Churchill meant, when the transform is applied to questions on electrical transients.

Problems

3.1 Prove (3.22).

3.2 Consider the following 'sort of similar-looking' time functions and show that, despite that 'similarity,' their Laplace transforms are quite different:

(a) $e^{-2t}u(t)$
(b) $e^{-2t}u(t-1)$
(c) $e^{-2(t-1)}u(t)$
(d) $e^{-2(t-1)}u(t-1)$

3.3 A time function that often occurs in electrical engineering is the *sine-integral*, defined as

$$S_i(t) = \int_0^t \frac{\sin(x)}{x} dx, \quad t \geq 0.$$

Calculate the Laplace transform of $S_i(t)$. Hint: Observe that $\frac{d}{dt}S_i(t) = \frac{\sin(t)}{t}$ (by invoking Leibniz's formula from note 3 in Chap. 1), and then use (3.16), (3.9), and (3.7).

3.4 For a and b both non-negative, what is the Laplace transform of

$$\frac{e^{-at} - e^{-bt}}{t}, t \geq 0, ?$$

Assume $a < b$. Hint: For the special case of $a = 0$ and $b = 1$, your answer should reduce to

$$\mathcal{L}\left\{\frac{1 - e^{-t}}{t}\right\} = \ln\left(1 + \frac{1}{s}\right).$$

Is your answer consistent with the initial and final value theorems? What happens if $a > b$?

---

[20] In a 1937 book that was the first to introduce the Laplace transform to engineers, the German mathematician Gustav Doetsch (1892–1977) was significantly less gracious toward Heaviside's operational calculus and its practitioners: see the Heaviside biography (note 9 in the Preface), p. 301. Heaviside had a running battle with academic mathematicians during his entire life. Even after election as a Fellow of the Royal Society, his submitted papers to the Society's *Proceedings* could be savaged. When that happened in 1894 (with the referee's report stating that Heaviside was "ignorant of the modern developments of the theory of linear differential equations"), he replied with these famous words: "[I admit] the rigorous logic of the matter is not plain! Well, what of that? *Shall I refuse my dinner because I do not fully understand the process of digestion* [my emphasis]? No, not if I am satisfied with the result." That, of course, failed to win the day with his mathematician critics.

3.5 Looking back at Fig. 1.1, and the defining voltage-current relations for a resistor, a capacitor, and an inductor, if we Laplace transform them under the assumption there is no initial charge or current then we get $V(s) = I(s)R$, $I(s) = CsV(s)$, and $V(s) = LsI(s)$, respectively. That is, in each case $I(s)$ and $V(s)$ are directly related through a function *only* of s, called the *generalized impedance* $Z(s) = \frac{V(s)}{I(s)}$: $Z_R(s) = R$, $Z_C(s) = \frac{1}{Cs}$, $Z_L(s) = sL$. This means we can treat *all three* circuit component types as 'generalized resistors,' and write the relationship between an applied voltage and a resulting current or voltage elsewhere in a circuit just as if *all* the components in the circuit are resistors.[21] Using this fact, show that the circuit of Fig. 3.14, for given values of L and C, behaves as a pure resistance for *any* $v(t)$ if $R = \sqrt{\frac{L}{C}}$. Further, in the circuit of Fig. 3.15, for any given values of L and R_1, for what values of R_2 and C does the circuit behave as a pure resistance for *any* $v(t)$?

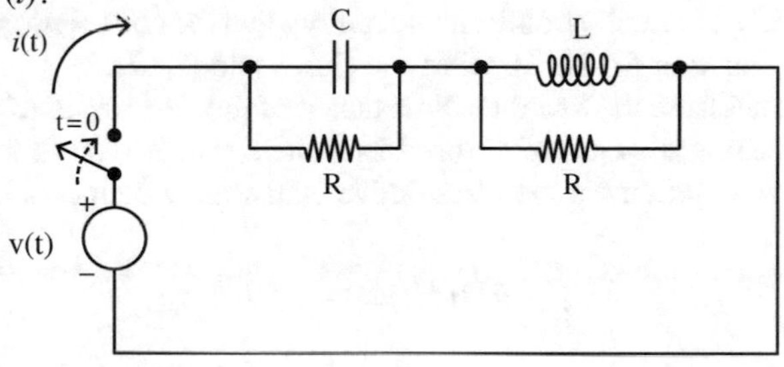

Fig. 3.14 Problem 3.5

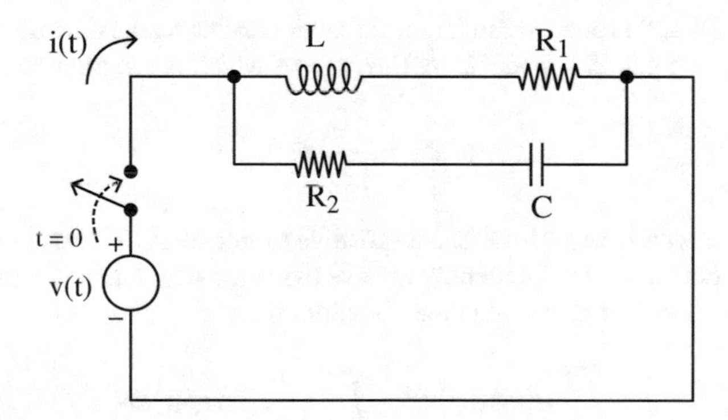

Fig. 3.15 Problem 3.5

[21]This was first explicitly stated by Oliver Heaviside in 1887 (he was using the *idea* as early as 1881). He called the expressions for Z_R, Z_C, and Z_L *resistance operators* (he was also the first to use both the symbol Z and the word *impedance* for a generalized resistance). For more on this, see the Heaviside biography (note 9 in the Preface, pp. 230–232). In Appendix 2 you'll find a discussion on how Heaviside used resistance operators to derive the required condition for distortionless signaling on an infinitely long transmission line. We'll use that condition, in Chap. 5, to find the transient behavior of such a line in response to a voltage step input.

3.6 Show that an impulse has infinite energy. Do that by applying the pulse (either as a voltage drop or as a current) of Fig. 3.8 to a resistor R, compute the energy delivered to R, and then let $\alpha \to 0$.

3.7 Using (3.7) and (3.45), what time function pairs with $\frac{1}{s^{5/2}}$? What time function has the Laplace transform $\frac{1}{1+\sqrt{s}}$? Hint: notice first that

$$\frac{1}{1+\sqrt{s}} = \frac{1}{1+\sqrt{s}} \cdot \frac{\sqrt{s}-1}{\sqrt{s}-1} = \frac{\sqrt{s}-1}{s-1} \cdot \frac{\sqrt{s}}{\sqrt{s}} = \frac{s-\sqrt{s}}{\sqrt{s}(s-1)} = \frac{s-1-\sqrt{s}+1}{\sqrt{s}(s-1)}$$
$$= \frac{s-1}{\sqrt{s}(s-1)} - \frac{\sqrt{s}}{\sqrt{s}(s-1)} + \frac{1}{\sqrt{s}(s-1)} = \frac{1}{\sqrt{s}} - \frac{1}{s-1} + \frac{1}{\sqrt{s}(s-1)},$$

and then realize that we have already derived the time functions that pair with each of these three final individual transforms.

3.8 Show, without actually doing an integration, that the convolution of an impulse with any function $f(t)$ is $f(t)$. Hint: use (3.52) and (3.87).

3.9 You'll recall from the text that Dirac talked of the *derivatives* of the impulse as being meaningful. One often useful result is that if $\phi(t)$ is a *differentiable* function (a requirement more restrictive than simply being continuous), then

$$\int_{-\infty}^{\infty} \delta'(t)\phi(t)dt = -\phi'(0).$$

See if you can develop a formal proof of this (that is, forget about being rigorous, just as I suspect did Dirac!). Hint: use integration-by-parts.

3.10 When Dirac introduced the impulse function to his readers, he mystified many with his so-called *impulse identities*, one of which is the often useful

$$\delta(at) = \frac{1}{|a|}\delta(t),$$

where a is any non-zero constant (positive *or* negative). What he meant by this is that both sides of the identity behave the same way *under the integral sign*. That is, if $\phi(t)$ is any continuous function, then

$$\int_{-\infty}^{\infty} \delta(at)\phi(t)dt = \int_{-\infty}^{\infty} \frac{1}{|a|}\delta(t)\phi(t)dt.$$

See if you can develop a formal proof. Hint: change variable in the integral on the left, and consider the two cases of $a > 0$ and $a < 0$.

3.11 If you found the previous problem pretty easy, here's a slightly more mysterious identity due to Dirac:

$$\int_{-\infty}^{\infty} \delta(u-x)\delta(x-v)dx = \delta(u-v).$$

Can you show this? Hint: with $\phi(t)$ as some continuous function, first evaluate the integral

$$\int_{-\infty}^{\infty} \phi(u)\left\{\int_{-\infty}^{\infty} \delta(u-x)\delta(x-v)dx\right\}du$$

by reversing the order of integration. Next, evaluate the integral

$$\int_{-\infty}^{\infty} \phi(u)\delta(u-v)du.$$

Notice anything interesting about your two results?

3.12 The convolution theorem tells us that if you multiply two transforms together, the result pairs with the convolution of the associated time functions. What, do you suppose, is the transform that pairs with the product of two time functions? That is, what goes on the right-hand-side of $f(t)g(t) \leftrightarrow$? Can you prove your answer? Hint: look into the Fourier transform, a close relative of the Laplace transform. We will *not* discuss the Fourier transform in this book, and so this problem is intended to be simply an open-ended challenge question for you to pursue on your own.[22]

3.13 In the text we developed the transform pairs

(a) $\frac{df}{dt} \leftrightarrow sF(s) - f(0+)$
(b) $u(t) \leftrightarrow \frac{1}{s}$
(c) $\delta(t) \leftrightarrow \frac{d}{dt}u(t) \leftrightarrow 1.$

Now, suppose $f(t) = u(t)$, as defined in (3.10). Then, (a) and (b) say

$$\frac{d}{dt}u(t) = \delta(t) \leftrightarrow s\left(\frac{1}{s}\right) - u(0+) = 1 - 1 = 0$$

[22] You can find more on Fourier transforms (including the answer to this problem) in my book *Dr. Euler's Fabulous Formula*, Princeton 2006, pp. 188–274.

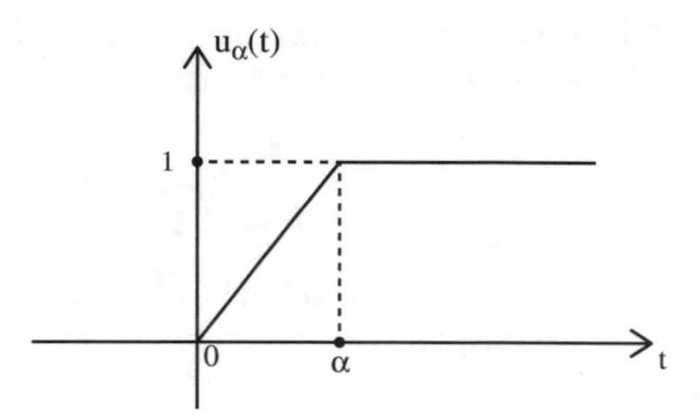

Fig. 3.16 Approximating the step as the limit of a *continuous* function

which doesn't agree with (c). Show that we can eliminate this conflict by re-defining the step function as shown in Fig. 3.16, which is *continuous* everywhere, even *at* $t = 0$, for any $\alpha \neq 0$. That is, by defining $u(t) = \lim_{\alpha \to 0} u_\alpha(t)$ where

$$u_\alpha(t) = \begin{matrix} 0, & t \leq 0 \\ \dfrac{t}{\alpha}, & 0 \leq t \leq \alpha \\ 1, & t \geq \alpha \end{matrix}$$

and so, for any $\alpha \neq 0$, $u_\alpha(0+) = 0+$. Note, carefully, that for the $f_\alpha(t)$ shown earlier in the text in Fig. 3.8, $f_\alpha(t) = \frac{d}{dt} u_\alpha(t)$.

Chapter 4
Transients in the Transform Domain

4.1 Voltage Surge on a Power Line

In this first example you'll see the full power of the Laplace transform in doing a traditional transient analysis. (You'll also experience its full grubbiness!) Figure 4.1 shows a circuit that is suddenly hit by a unit step voltage $v(t) = u(t)$, and our problem is to determine the resulting voltage $e(t)$. This circuit might, for example, be a simple model for a power station transformer connected (through the 30 μH inductor) to an overhead transmission line that has just been hit by a lightning stroke, a potentially catastrophic event (simply scale our final result up from our assumed one-volt surge to, say, a more realistic 100,000 volts). Determining $e(t)$ would tell the transformer designers what sort of 'safety-factor' they should consider for the survival of the transformer when confronted by such a large voltage surge, both in terms of the magnitude and the duration of the surge.

Going directly to transforms, Kirchhoff's current law at the $e(t)$ node is (remember, using microhenrys, microfarads, and ohms gives time in microseconds)

$$\frac{V-E}{30s} - \frac{E}{30} + E(0.05s) + \frac{E}{60s + \frac{1}{0.033s}}.$$

With some algebra, this becomes

$$E = \left(\frac{V}{30}\right) \frac{11.88s^2 + 6}{0.594s^4 + 0.396s^3 + 0.894s^2 + 0.2s + 0.2}. \tag{4.1}$$

Dividing the two polynomials by 6, and then multiplying through the denominator polynomial by the 30, (4.1) becomes (after replacing V with $\frac{1}{s}$, the transform of the unit step)

P. J. Nahin, *Transients for Electrical Engineers*,
https://doi.org/10.1007/978-3-319-77598-2_4

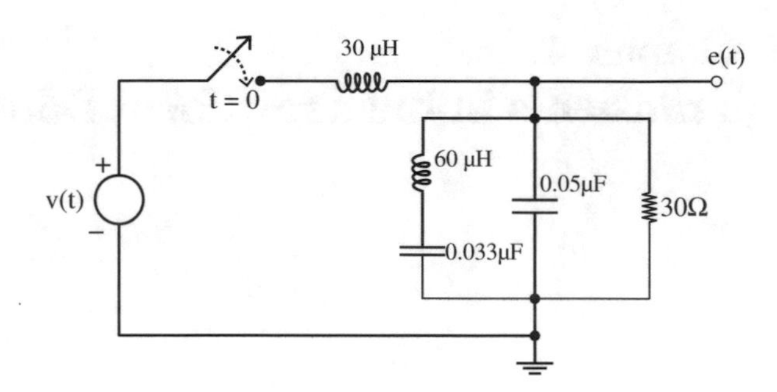

Fig. 4.1 Power station transformer model

$$E = \frac{1.98s^2 + 1}{s(2.97s^4 + 1.98s^3 + 4.47s^2 + s + 1)}$$

or,

$$E = \frac{1.98s^2 + 1}{2.97s(s^4 + 0.667s^3 + 1.505s^2 + 0.337s + 0.337)}. \tag{4.2}$$

At this point we could simply plug the ratio of polynomials in (4.2) into a suitably modified **pulse.m**, the code given at the end of the last chapter — and we *will* do that at the end of this analysis — but for now let's continue-on with the formal, traditional approach.

Because we know we will have to do a partial fraction expansion of (4.2), we now need to factor the denominator polynomial, a miserable task once performed in yesteryears by complicated algebraic techniques. Since this is the twenty-first century we will instead invoke a computer software package to do that job for us.[1] The result is that the roots of

$$s^4 + 0.667s^3 + 1.505s^2 + 0.337s + 0.337 = 0$$

are the two conjugate pairs

$$-0.083 \pm i0.542 \text{ and } -0.25 \pm i1.029.$$

Thus,

$$s^4 + 0.667s^3 + 1.505s^2 + 0.337s + 0.337$$

[1]In *MATLAB*, for example, all that one does is type

$$solve('x\text{^}4 + 0.667*x\text{^}3 + 1.505*x\text{^}2 + 0.337*x + 0.337')$$

and the roots almost instantly appear on the computer screen.

$$= (s + 0.083 + i0.542)(s + 0.083 - i0.542)(s + 0.25 + i1.1029) \\ \times (s + 0.25 - i1.1029) \\ = \left[(s + 0.083)^2 + (0.542)^2\right]\left[(s + 0.25)^2 + (1.029)^2\right] \\ = \left(s^2 + 0.166s + 0.301\right)\left(s^2 + 0.5s + 1.121\right)$$

and so

$$E = \frac{1.98s^2 + 1}{2.97s(s^2 + 0.166s + 0.301)(s^2 + 0.5s + 1.121)} \tag{4.3}$$

which we can write as the partial fraction expansion

$$E = \frac{As + B}{s^2 + 0.5s + 1.121} + \frac{Cs + D}{s^2 + 0.166s + 0.301} + \frac{F}{s}. \tag{4.4}$$

The value of F is easy to find: set (4.4) equal to (4.3), multiply through by s, and set $s = 0$ to get

$$F = \frac{1}{(2.97)(0.301)(1.121)} = 0.9979.$$

Actually, we know that $F = 1$, *exactly*, since the circuit physics[2] tells us that

$$\lim_{t \to \infty} e(t) = 1,$$

and the discrepancy of 0.0021 in the value of F (an error of less than $\frac{1}{4}$ of 1%) is simply round-off error in the coefficients of (4.3). We'll use $F = 1$, not 0.9979.

To find the other constants in the partial fraction expansion, multiply through (4.3) set equal to (4.4) by the denominator in (4.3) to get

$$1.98s^2 + 1 = 2.97s\left(s^2 + 0.166s + 0.301\right)(As + B) \\ + 2.97s\left(s^2 + 0.5s + 1.121\right)(Cs + D) \\ + 2.97\left(s^2 + 0.5s + 1.121\right)\left(s^2 + 0.166s + 0.301\right)F.$$

Then, equating the coefficients of equal powers of s:

$$s^1 : 0 = 2.97[0.301B + 1.121D + (0.5)(0.301)F + (1.121)(0.166)F]$$

or, using $F = 1$,

$$0.301B + 1.121F = -0.337. \tag{4.5}$$

[2]In the d-c steady state, the inductors have zero voltage drop and the capacitors are open-circuits. The entire input, as $t \to \infty$, appears across the 30 ohm resistor.

$$s^2 : 1.98$$
$$= 2.97[0.166B + 0.301A + 0.5D + 1.121C + 1.121F + 0.301F + (0.5)(0.166)F]$$

or, using $F = 1$,

$$0.301A + 0.166B + 1.121C + 0.5D = -0.838. \tag{4.6}$$
$$s^3 : 0 = 2.97[0.166A + B + 0.5C + D + 0.5F + 0.166F]$$

or, using $F = 1$,

$$0.166A + B + 0.5C + D = -0.666. \tag{4.7}$$
$$s^4 : 0 = 2.97[A + C + F]$$

or, using $F = 1$,

$$A + C = -1. \tag{4.8}$$

The equations (4.5) through (4.8) are four simultaneous algebraic equations in four unknowns, and most easily solved by writing them in matrix form as

$$\boldsymbol{M}\begin{bmatrix} A \\ B \\ C \\ D \end{bmatrix} = \begin{bmatrix} -0.337 \\ -0.838 \\ -0.666 \\ -1 \end{bmatrix}$$

where

$$\boldsymbol{M} = \begin{bmatrix} 0 & 0.301 & 0 & 1.121 \\ 0.301 & 0.166 & 1.121 & 0.5 \\ 0.166 & 1 & 0.5 & 1 \\ 1 & 0 & 1 & 0 \end{bmatrix}.$$

We easily[3] solve for A, B, C, and D by computing

[3]I'm being just a bit flippant here: I use the word *easily* because *MATLAB* did all the work. After entering $\boldsymbol{M}$ and then the column vector $\boldsymbol{b} = \begin{bmatrix} -0.337 \\ -0.838 \\ -0.666 \\ -1 \end{bmatrix}$, all I did was type *inv*($\boldsymbol{M}$) $*$ $\boldsymbol{b}$ and the almost instant result was the solution column vector $\begin{bmatrix} -0.531 \\ -0.058 \\ -0.469 \\ -0.285 \end{bmatrix}$.

$$\begin{bmatrix} A \\ B \\ C \\ D \end{bmatrix} = \boldsymbol{M}^{-1} \begin{bmatrix} -0.337 \\ -0.838 \\ -0.666 \\ -1 \end{bmatrix} = \begin{bmatrix} -0.531 \\ -0.058 \\ -0.469 \\ -0.285 \end{bmatrix}.$$

Thus,

$$E(s) = -\frac{0.531s + 0.058}{s^2 + 0.5s + 1.121} - \frac{0.469s + 0.285}{s^2 + 0.166s + 0.301} + \frac{1}{s}. \tag{4.9}$$

We physically expect to see exponentially damped sinusoidal oscillations (why?), and so anticipate a need for the following transform pairs (which follow from (3.13), (3.15), and (3.16)):

$$e^{-\alpha t} \cos(\omega t) \leftrightarrow \frac{s + \alpha}{(s + \alpha)^2 + \omega^2} \tag{4.10}$$

and

$$e^{-\alpha t} \frac{\sin(\omega t)}{\omega} \leftrightarrow \frac{1}{(s + \alpha)^2 + \omega^2}. \tag{4.11}$$

We can get the quadratic denominators of $E(s)$ in (4.9) into the form of these transforms by writing

$$s^2 + as + b = (s + \alpha)^2 + \omega^2 = s^2 + 2\alpha s + \alpha^2 + \omega^2$$

and thus

$$2\alpha = a, \text{or } \alpha = \frac{1}{2}a$$

and

$$\alpha^2 + \omega^2 = b, \text{or } = \sqrt{b - \alpha^2} = \sqrt{b - \frac{1}{4}a^2}.$$

So,

$$s^2 + 0.5s + 1.121 = (s + 0.25)^2 + (1.029)^2$$

and

$$s^2 + 0.166s + 0.301 = (s + 0.083)^2 + (0.542)^2.$$

Okay, now (finally!) we are in the homestretch. Using these last results in (4.9), we have

$$E(s) = -\frac{0.531s + 0.058}{(s+0.25)^2 + (1.029)^2} - \frac{0.469s + 0.285}{(s+0.083)^2 + (0.542)^2} + \frac{1}{s}$$

$$= -0.531\left[\frac{s+0.109}{(s+0.25)^2 + (1.029)^2}\right] - 0.469\left[\frac{s+0.608}{(s+0.083)^2 + (0.542)^2}\right] + \frac{1}{s}$$

$$= -0.531\left[\frac{s+0.25}{(s+0.25)^2 + (1.029)^2} - \frac{0.25-0.109}{(s+0.25)^2 + (1.029)^2}\right]$$

$$-0.469\left[\frac{s+0.083}{(s+0.083)^2 + (0.542)^2} + \frac{0.608-0.083}{(s+0.083)^2 + (0.542)^2}\right] + \frac{1}{s}$$

$$= -0.531\left[\frac{s+0.25}{(s+0.25)^2 + (1.029)^2} - \frac{0.141}{(s+0.25)^2 + (1.029)^2}\right]$$

$$-0.469\left[\frac{s+0.083}{(s+0.083)^2 + (0.542)^2} + \frac{0.525}{(s+0.083)^2 + (0.542)^2}\right] + \frac{1}{s}$$

or, at last,

$$E(s) = -0.531\frac{s+0.25}{(s+0.25)^2 + (1.029)^2} + 0.075\frac{1}{(s+0.25)^2 + (1.029)^2}$$
$$-0.469\frac{s+0.083}{(s+0.083)^2 + (0.542)^2} - 0.246\frac{1}{(s+0.083)^2 + (0.542)^2} + \frac{1}{s}. \tag{4.12}$$

Using the transform pairs in (4.10) and (4.11), we can now immediately write down $e(t)$ from (4.12):

$$e(t) = -0.531e^{-0.25t}\cos(1.029t) + 0.075\frac{e^{-0.25t}\sin(1.029t)}{1.029}$$
$$-0.469e^{-0.083t}\cos(0.542t) - 0.246\frac{e^{-0.083t}\sin(0.542t)}{0.542} + 1$$
$$= -0.531e^{-0.25t}\cos(1.029t) + 0.073e^{-0.25t}\sin(1.029t)$$
$$-0.469e^{-0.083t}\cos(0.542t) - 0.454e^{-0.083t}\sin(0.542t) + 1$$

or, *at last*,

$$e(t) = \begin{bmatrix} 1 - e^{-0.25t}\{0.531\cos(1.029t) - 0.073\sin(1.029t)\} \\ -e^{-0.083t}\{0.469\cos(0.542t) + 0.454\sin(0.542t)\} \end{bmatrix} u(t). \tag{4.13}$$

Wow, what a calculation!

Figure 4.2 shows a plot of (4.13) for the first 30 microseconds of $e(t)$, which reaches a peak of about 1.35 times the surge amplitude in just under 10 microseconds after the start of the unit step surge. As expected, the response approaches 1 as $t \to \infty$.

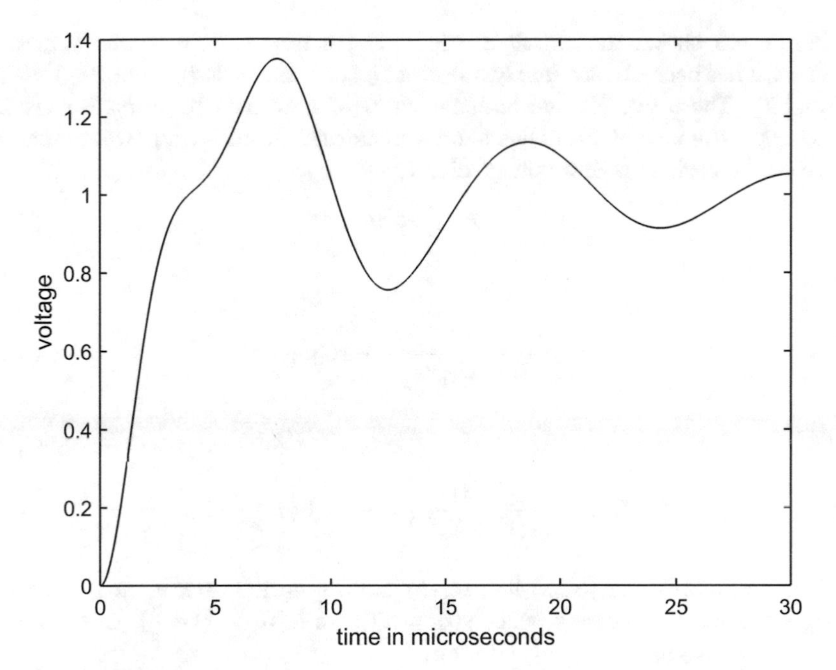

Fig. 4.2 The surge response of the circuit in Fig. 4.1

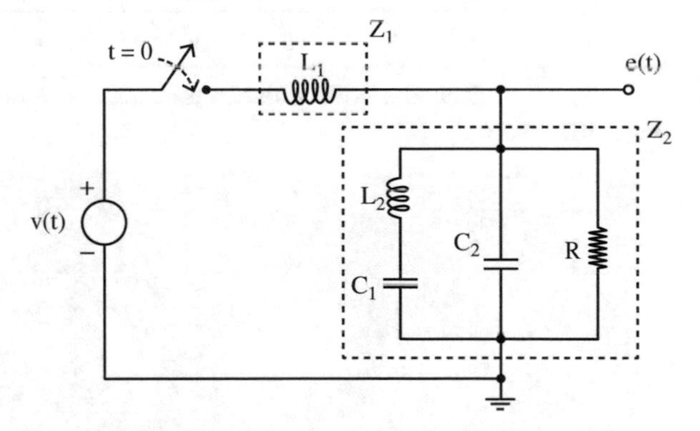

Fig. 4.3 A generalization of Fig. 4.1

Now, suppose the pleasure at having successfully slogged your way through this analysis is suddenly soaked with a bucket of cold water when you are told that, so sorry, but somebody mis-read the numbers and accidently reversed the values of the two inductors and two capacitors. In addition, that 30 ohm resistor should have been 3.0 ohms. Would you mind terribly re-doing the analysis? If it's your boss asking, well then, *of course* you'll quickly reply 'Hey, no problem!' But we all know what you'll be *thinking*! But now, let's do it the smart way, the modern way, using a combination of Heaviside's resistance operators (discussed in Problem 3.5 of the previous chapter) and a computer.

Figure 4.3 shows the circuit of Fig. 4.1, but now with symbolic values. In addition, it has been divided into two sections (each inside a dashed-line box) labeled Z_1 and Z_2. These two Z's are both functions of s, formed by using Heaviside's resistor operator insight. From this figure we can write the following two expressions based on the basic idea of a voltage divider:

$$Z_1(s) = sL_1 \tag{4.14}$$

and

$$\frac{1}{Z_2(s)} = \frac{1}{sL_2 + \frac{1}{sC_1}} + sC_2 + \frac{1}{R} \tag{4.15}$$

where

$$E(s) = \frac{Z_2(s)}{Z_1(s) + Z_2(s)} V(s) = V(s)\frac{1}{\frac{Z_1}{Z_2}} + 1. \tag{4.16}$$

Plugging (4.14) and (4.15) into (4.16), and doing the three minutes of pretty straightforward algebra (and, since $v(t) = u(t)$, we have $V(s) = \frac{1}{s}$), it is not at all difficult to arrive at

$$E(s) = \frac{s^2 L_2 C_1 + 1}{s\left[s^4 L_1 L_2 C_1 C_2 + s^3 \frac{L_1 L_2 C_1}{R} + s^2(L_1 C_1 + L_2 C_1 + L_1 C_2) + s\frac{L_1}{R} + 1\right]}. \tag{4.17}$$

Now, go back to the code **pulse.m** and make the obvious changes to arrive at **transformer.m**:

```
%transformer.m/created by PJNahin for Electrical Transients
(7/17/2017)
L1=30;L2=60;C1=.033;C2=.05;R=30;
k1=L2*C1;k2=L1*L2*C1*C2;k3=L1*L2*C1/R;k4=L1*C1+L2*C1
+L1*C2;k5=L1/R;
syms s t
h=ilaplace((k1*s^2+1)/(s*(k2*s^4+k3*s^3+k4*s^2+k5*s
+1)));
t=linspace(0,30,250);
y=subs(h);
v=vpa(y);
plot(t,v,'-k')
xlabel('time in microseconds')
ylabel('voltage')
```

We can partially check (4.17) by first setting the component values to the original numbers (as is done in **transformer.m**), and then watch the code produce Fig. 4.4, which is virtually identical to Fig. 4.2. It is then a task of mere seconds to re-do the problem with the 'corrected' values of $L_1 = 60$, $L_2 = 30$, $C_1 = 0.05$, $C_2 = 0.033$, and

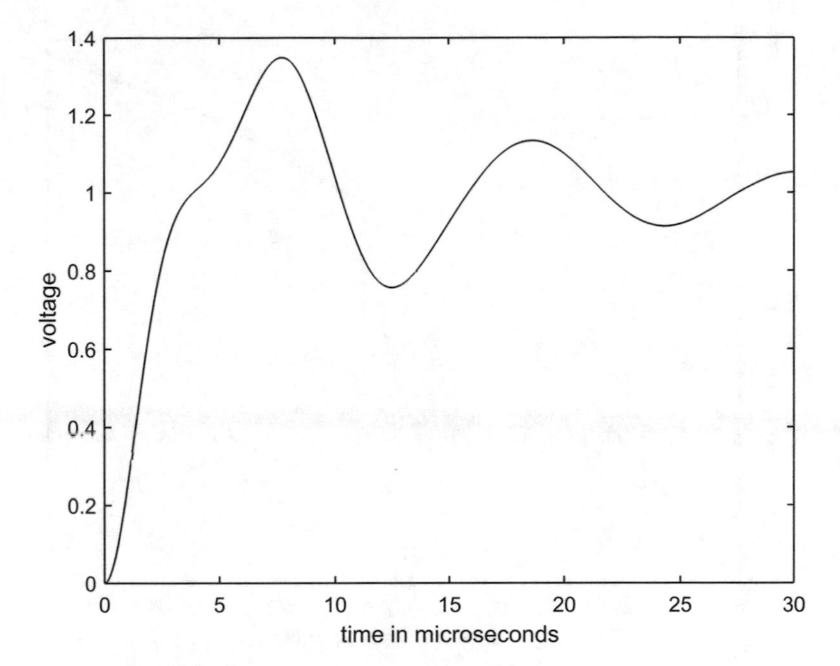

Fig. 4.4 Compare with Fig. 4.2

$R = 3$, with the result being Fig. 4.5. *Now* if the boss comes back with even more new values for the components, there really isn't any problem at all (assuming, of course, that you have access to a computer running the right software).

4.2 Two Hard Problems from Yesteryear

Here's another example of how the computer has changed transient analysis. Figure 4.6 shows a circuit I've taken from a 1935 textbook (note 16 in the Preface), a circuit which the authors solved using a *hugely* laborious process very much like the first part of the last section — only worse! In particular, the problem is to find the battery current *i*(*t*) after the switch is closed at $t = 0$, assuming there is no energy initially stored in the circuit.

The equations describing the circuit are

$$Z_1(s) = R + sL, \tag{4.18}$$

$$Z_2(s) = R_1 + sL_1, \tag{4.19}$$

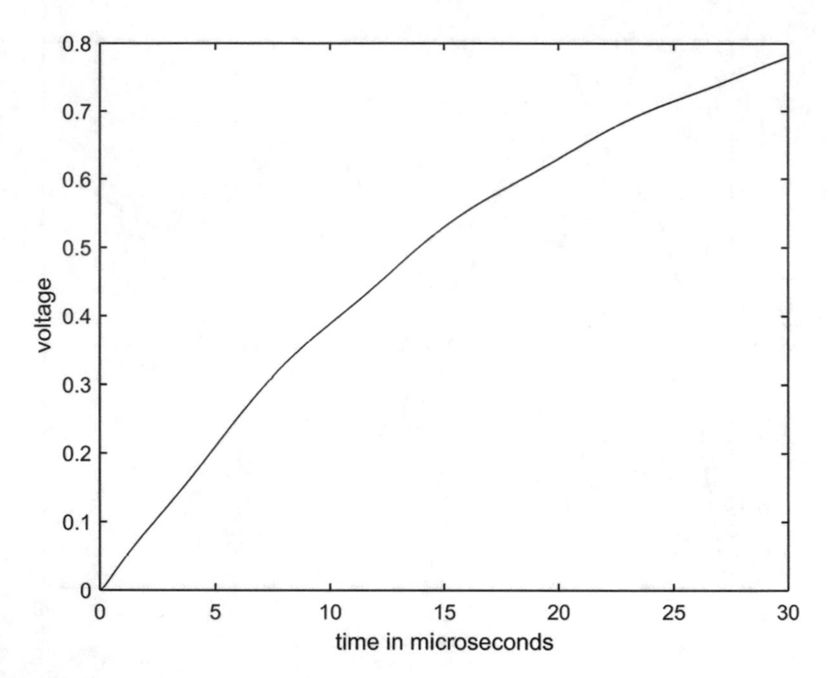

Fig. 4.5 'Corrected' response

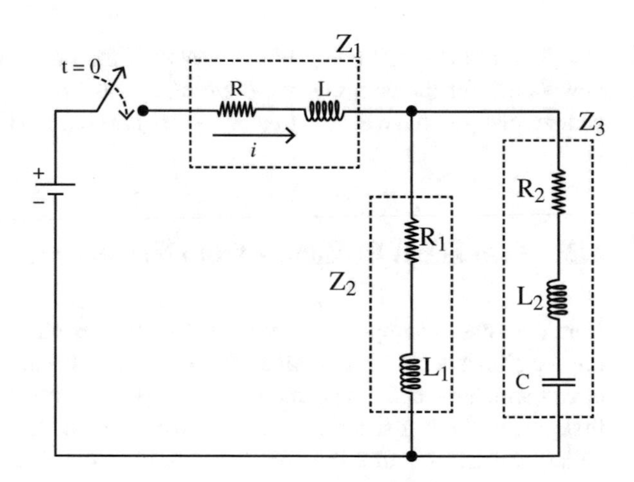

Fig. 4.6 What is the battery current $i(t)$?

and

$$Z_3(s) = R_2 + sL_2 + \frac{1}{sC}. \tag{4.20}$$

Then, since Z_1 is in series with the parallel combination of Z_2 and Z_3, we have the transform of the current as

$$I(s) = \frac{V(s)}{Z_1 + \frac{Z_2 Z_3}{Z_2 + Z_3}}$$

or,

$$I(s) = V(s) \frac{Z_2 + Z_3}{Z_1 Z_2 + Z_1 Z_3 + Z_2 Z_3}. \tag{4.21}$$

Inserting (4.18), (4.19), and (4.20) into (4.21), and being careful with the algebra, you should be able to show that

$$I(s) = V(s) \frac{s^2(L_1 + L_2)C + s(R_1 + R_2)C + 1}{D(s)} \tag{4.22}$$

where

$$\begin{aligned} D(s) = {} & s^3(LL_1C + LL_2C + L_1L_2C) \\ & + s^2(R_1CL + RCL_1 + L_2RC + R_2CL + R_1CL_2 + L_1R_2C) \\ & + s(RR_1C + RR_2C + L + R_1R_2C + L_1) + R + R_1. \end{aligned}$$

If we put (4.22) into our basic code from the end of Chap. 3, then all we have left to do is declare the values of the components (along with the actual battery voltage). I'll use the values given in the 1935 textbook, values that may look just a bit chaotic at first: $R = 1$ ohm, $R_1 = 17.7$ ohms, $R_2 = 5$ ohms, $L = 0.011\ H$, $L_1 = 0.112\ H$, $L_2 = 0.056\ H$, and $C = 28.8\ \mu F$, along with a 30.5 volt battery. These are the values[4] in the code **KC1.m**, for the $I(s)$ given in (4.22). (The code's name is in honor of the authors of that 1935 textbook, Professors Kurtz and Corcoran.) When run, **KC1.m** produced Fig. 4.7, a plot that required the support of *five pages* in the textbook of really mind-numbing arithmetic. And every time any of the component values change, another five pages of agony would be required, while the code only needs to have its first line retyped.

[4]With the component values in ohms, millihenrys, and millifarads, time will be measured in milliseconds. So, $L = 11$, $L_1 = 112$, $L_2 = 56$, and $C = 0.0288$. Now, *why* these curious values? My guess is that Kurtz and Corcoran just used whatever components were available in a spare-parts box on a shelf in their lab (they actually built the circuit of Fig. 4.6 and included a strip-chart recording of $i(t)$ in their book) and measured the values after-the-fact. By the way, their experimental strip-chart recording looks *exactly* like the computer-generated Fig. 4.7.

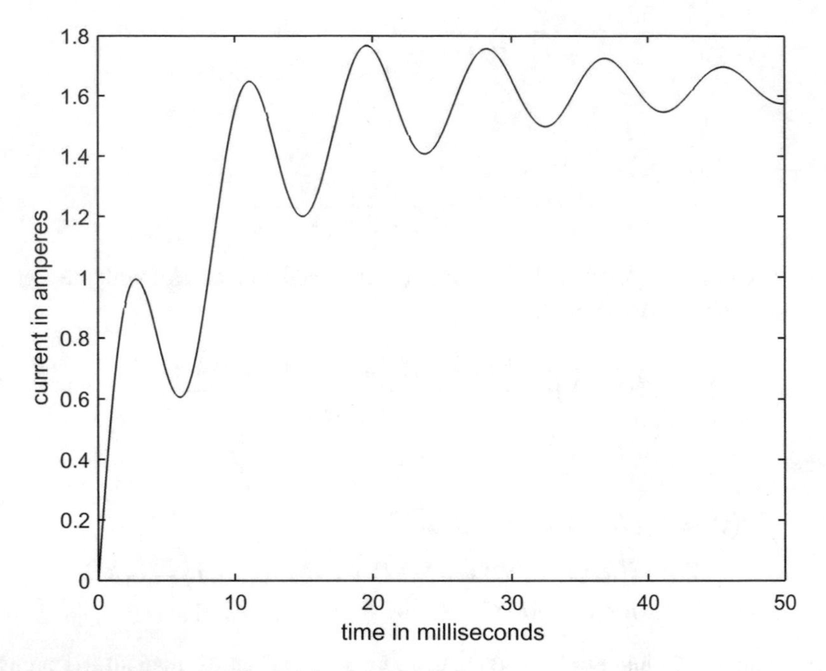

Fig. 4.7 The battery current for the circuit in Fig. 4.6

```
%KC1.m/created by PJNahin for Electrical Transients (7/18/
2017)
R=1;L=11;R1=17.7;L1=112;R2=5;L2=56;C=.0288;
k1=(L1+L2)*C;k2=(R1+R2)*C;k3=L*L1*C+L*L2*C+L1*L2*C;
k4=R1*C*L+R*C*L1+L2*R*C+R2*C*L+R1*C*L2+L1*R2*C;
k5=R*R1*C+R*R2*C+L+R1*R2*C+L1;k6=R+R1;
syms s t
h=ilaplace((k1*s^2+k2*s+1)/(s*(k3*s^3+k4*s^2+k5*s+k6)));
h=30.5*h;
t=linspace(0,50,250);
y=subs(h);
v=vpa(y);
plot(t,v,'-k')
xlabel('time in milliseconds')
ylabel('current in amperes')
```

The reason the code is called **KC1.m** is because there is a **KC2.m**, a code that solves another problem from that same textbook. That circuit, the one we are going to study next, is shown in Fig. 4.8, and it involves a subject we touched-on briefly at the end of Chap. 1 and in Chap. 2 — magnetic coupling between different inductors in the circuit. There we discussed the impact of magnetic coupling on the natural frequency of a tuned pair of circuits. Our problem here is to find the actual transient primary and secondary currents $i_1(t)$ and $i_2(t)$, respectively, for $t \geq 0$ when the switch

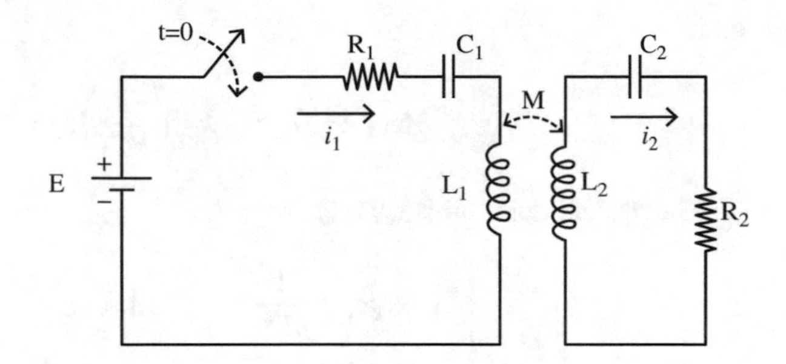

Fig. 4.8 Magnetically coupled, *un*-tuned circuit

is closed at $t = 0$ when (unlike in Chap. 2) the coupled circuits *have energy loss*. The component values are $R_1 = 3.5$ ohms, $R_2 = 0.8$ ohms, $L_1 = 0.093$ H, $L_2 = 0.011$ H, $C_1 = 150\ \mu$F, and $C_2 = 168\ \mu$F.

The Kirchhoff voltage loop law equations for Fig. 4.8 are

$$L_1 \frac{di_1}{dt} + R_1 i_1 + \frac{1}{C_1} \int_0^t i_1(x)dx \pm M \frac{di_2}{dt} = E \tag{4.23}$$

and

$$L_2 \frac{di_2}{dt} + \frac{1}{C_2} \int_0^t i_2(x)dx + R_2 i_2 \pm M \frac{di_1}{dt} = 0 \tag{4.24}$$

where M is a *positive* number. Recall from Chap. 1 why the plus-minus signs are in front of M in (4.23) and (4.24): because the polarity of the induced voltage due the mutual inductance in both the primary and secondary circuits can indeed have either sign, depending on the relative sense of the windings of the coils in the individual inductors. Kurtz and Corcoran used the plus sign in both circuits, and we'll follow their lead.

So, transforming (4.23) and (4.24),

$$L_1 s I_1 + R_1 I_1 + \frac{1}{sC_1} I_1 + M s I_2 = \frac{E}{s}$$

and

$$L_2 s I_2 + \frac{1}{sC_2} I_2 + R_2 I_2 + M s I_1 = 0.$$

Re-writing these last two equations in the form for Cramer's rule,

$$\left(L_1 s + R_1 + \frac{1}{sC_1} \right) I_1 + M s I_2 = \frac{E}{s} \tag{4.25}$$

and

$$MsI_1 + \left(L_2 s + R_2 + \frac{1}{sC_2}\right) I_2 = 0. \tag{4.26}$$

The system determinant, D, is

$$D = \begin{vmatrix} L_1 s + R_1 + \dfrac{1}{sC_1} & Ms \\ Ms & L_2 s + R_2 + \dfrac{1}{sC_2} \end{vmatrix}$$
$$= \left(L_1 s + R_1 + \frac{1}{sC_1}\right)\left(L_2 s + R_2 + \frac{1}{sC_2}\right) - M^2 s^2$$

or,

$$D = \left(\frac{C_1 L_1 s^2 + R_1 C_1 s + 1}{sC_1}\right)\left(\frac{C_2 L_2 s^2 + R_2 C_2 s + 1}{sC_2}\right) - M^2 s^2. \tag{4.27}$$

From Cramer's rule,

$$I_1 = \frac{\begin{vmatrix} \dfrac{E}{s} & Ms \\ 0 & L_2 s + R_2 + \dfrac{1}{sC_2} \end{vmatrix}}{D}$$

and

$$I_2 = \frac{\begin{vmatrix} L_1 s + R_1 + \dfrac{1}{sC_1} & \dfrac{E}{s} \\ Ms & 0 \end{vmatrix}}{D}.$$

Doing the algebra, we soon have

$$I_1(s) = E \frac{s^2 L_2 C_1 C_2 + sR_2 C_1 C_2 + C_2}{k_3 s^4 + k_4 s^3 + k_5 s^2 + k_6 s + 1} \tag{4.28}$$

where

$$k_3 = C_1 L_1 C_2 L_2 - M^2 C_1 C_2, k_4 = C_1 L_1 R_2 C_2 + R_1 C_1 C_2 L_2,$$
$$k_5 = C_1 L_1 + R_1 C_1 R_2 C_2 + C_2 L_2, k_6 = R_1 C_1 + R_2 C_2.$$

Also,

$$I_2(s) = -EM \frac{s^2 C_1 C_2}{k_3 s^4 + k_4 s^3 + k_5 s^2 + k_6 s + 1}. \tag{4.29}$$

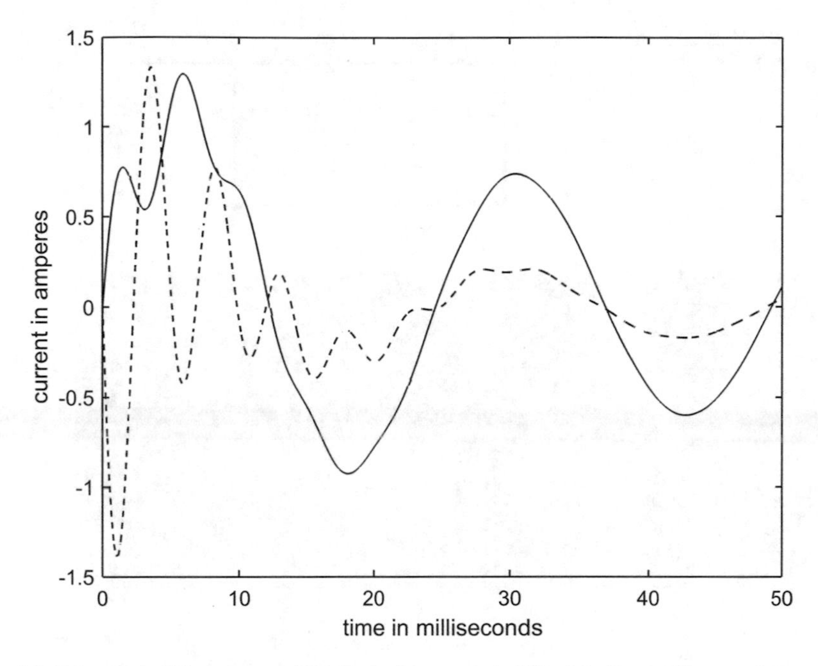

Fig. 4.9 Primary (solid) and secondary (dashed) currents in Fig. 4.8. (Again, Kurtz and Corcoran actually constructed the circuit in Fig. 4.8 and included strip-chart recordings of the primary and secondary currents in their book. The computer-generated curves in Fig. 4.9 are extremely close to the experimental recordings.)

The code **KC2.m** carries out the inverse transform evaluation of (4.28) and (4.29) over the first 50 milliseconds after the switch is closed, and generates the plot shown in Fig. 4.9. In that figure the solid curve is the primary current, while the dashed curve is the secondary current. As those curves show, both transient currents look noticeably non-sinusoidal, a feature directly due to the energy loss mechanisms (R_1 and R_2) in the primary and secondary circuits.

```
%KC2.m/created by PJNahin for Electrical Transients (7/19/
2017)
E=30;R1=3.5;R2=0.8;L1=93;L2=11;M=26;C1=0.15;C2=0.168;
k1=L2*C1*C2;k2=R2*C1*C2;k3=C1*L1*C2*L2-M*M*C1*C2;
k4=C1*L1*R2*C2+R1*C1*C2*L2;
k5=C1*L1+R1*C1*R2*C2+C2*L2;k6=R1*C1+R2*C2;
syms s t
h=ilaplace((k1*s^2+k2*s+C2)/(k3*s^4+k4*s^3+k5*s^2+k6*s
+1));
h=E*h;
t=linspace(0,50,300);
y=subs(h);
v=vpa(y);
```

(continued)

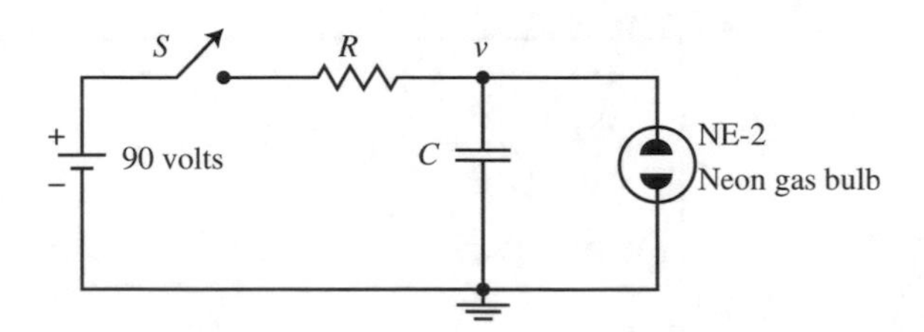

Fig. 4.10 What does this circuit do?

```
plot(t,v,'-k')
hold on
k7=C1*C2;
h=ilaplace((k7*s^2)/(k3*s^4+k4*s^3+k5*s^2+k6*s+1));
h=-E*M*h;
y=subs(h);
v=vpa(y);
plot(t,v,'--k')
xlabel('time in milliseconds')
ylabel('current in amperes'
```

4.3 Gas-Tube Oscillators

In this problem we return to a direct use of the Laplace transform, without any computer involvement, to study a circuit with a new (for this book) type of electrical component: the neon gas bulb.[5] A simple and easy to understand circuit using such a gas bulb is shown in Fig. 4.10, where we suppose that the bulb is initially 'off,' that is, not conducting current. Then, we close the switch S at time $t = 0$. What happens next?

(1) If the bulb voltage drop (v) across the bulb terminals is less than some critical voltage V_s (and it will be at time $t = 0$ if we assume that C is initially uncharged), then the gas is a non-conductor and the bulb remains an open circuit.
(2) The capacitor C begins to be charged by the 90-volt battery through the resistor R and so the capacitor voltage (that is, v) rises.
(3) When the capacitor voltage v reaches V_s the gas ionizes or *strikes* (that is, the electric field inside the bulb has become large enough that the valence electrons in the neon atoms are ripped-free from their orbits and the gas becomes a mix of

[5]In this section we will be discussing, in particular, the NE-2 gas bulb, which is about the size of a finger-tip (perhaps a bit smaller), and available today for about 15 cents each if bought in lots of ten or more. It has been around for decades (I remember building the NE-2 circuits discussed in this section in the mid-1950s when I was in high school).

free negative charges and positive ions). The free electrons therefore possess, temporarily, extra energy. The gas itself is now a very good conductor.

(4) Once ionized, the now highly conductive gas presents a low-resistance path across the capacitor, and so the capacitor *very rapidly* dumps some of its charge through the bulb. The charge dump continues until v drops below some critical voltage $V_E < V_s$, at which point the gas in the bulb recombines and the bulb returns to being a non-conductor.

(5) The recombination of the previously ionized gas means that the extra energy once possessed by the free electrons (now no longer free) is given up as radiation at a wavelength characteristic of the gas (reddish-yellow for neon: blue if the bulb gas was argon, instead). That is, the bulb emits a flash of visible light.

(6) The circuit is now back to its original state, and the entire process begins anew. The visual result is a periodically blinking bulb, at a rate that is easily calculated, and what follows is how to do that.

After the gas recombines we have $v = V_E$, with the capacitor recharging back towards 90 volts. When v reaches V_S the gas ionizes again . . . now read (3) — (6). In summary, the bulb is off and the capacitor is recharging during the time it takes for the capacitor to exponentially charge from $v = V_E$ to $v = V_S$ with a time constant of RC and a battery voltage of 90 volts. At $v = V_S$ the bulb turns on, v quickly falls to V_E, the bulb flashes, and it all starts over again. The general equation for v while C is charging is

$$v = 90\left(1 - e^{-t/RC}\right),$$

and what we want to know is how long it takes v to go from V_E to V_S. Let $t = T_1$ and $t = T_2$ be such that

$$v(T_1) = 90\left(1 - e^{-T_1/RC}\right) = V_E$$

and

$$v(T_2) = 90\left(1 - e^{-T_2/RC}\right) = V_S.$$

The time duration we are interested in, then, is (after a bit of easy algebra) given by

$$\Delta t = T_2 - T_1 = RCln\left(\frac{90 - V_E}{90 - V_S}\right). \tag{4.30}$$

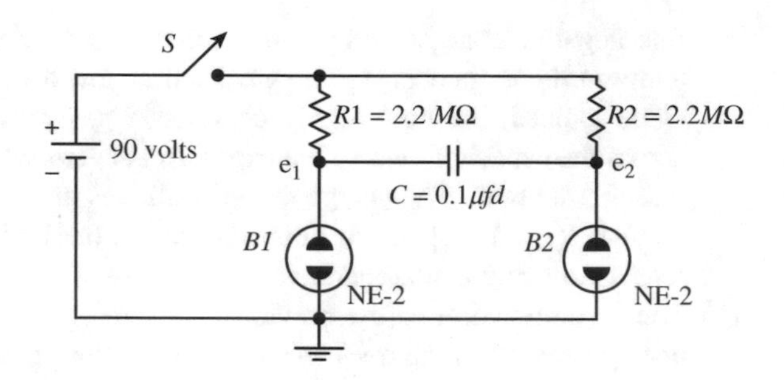

Fig. 4.11 What does this circuit do?

No two gas bulbs are identical (it is virtually certain that two randomly picked bulbs will have different values of V_S and V_E), but for the NE-2 the values are typically somewhere in the intervals 70 volts < V_S < 80 volts and 50 volts < V_E < 55 volts. Suppose, for example, that $V_E = 52$ volts and $V_S = 75$ volts. If $RC = 1$ second (say, $R = 1$ megohm and $C = 1$ microfarad) then

$$\Delta t = \ln\left(\frac{90-52}{90-75}\right) = 0.93 \text{ seconds.}$$

That is, the NE-2 bulb in the circuit of Fig. 4.10 would flash about 65 times per minute. This circuit, called a *relaxation oscillator*, is in a *never-ending* state of transient behavior.

This was all pretty straightforward, but matters are somewhat less obvious for the NE-2 circuit of Fig. 4.11. Now we have *two* bulbs, and neither one has a capacitor directly across it as in Fig. 4.10. The claim is that the pair of bulbs *alternately* turn on-and-off at a steady rate; our problem is to show that this is so, as well as to calculate the on-off rate for each bulb.

To understand the behavior of the pair of bulbs in Fig. 4.11, you need to have a slightly refined idea on what happens after a gas tube strikes. Once ionized, the gas doesn't simply become a low-resistance path (or, in the most extreme version, a 'short'), but rather it maintains a constant voltage drop of V_C (where $V_E < V_C < V_S$) across the bulb terminals.[6] As long as the external circuitry is capable of providing a voltage drop of V_C across the bulb, then the external circuitry will 'absorb' any excess voltage in whatever way necessary to be consistent with Kirchhoff's laws. If the external circuitry cannot support a voltage drop of at least V_E across the bulb, however, then the bulb will turn off and become a non-conductor once again. When a capacitor is connected directly across a bulb, as in Fig. 4.10, this refinement really

[6]The constant voltage drop behavior of an ionized gas tube once made it a favorite of designers of voltage-regulated power supplies, and of other circuitry needing a fixed reference voltage. A battery could do that, of course, but batteries eventually grow old, die, and need to be replaced. What is really wanted is an *automatic* reference voltage source *guaranteed* to *always* be available, and the gas tube filled the bill. It has, however, been replaced in modern circuitry with the solid-state Zener diode, which can be manufactured to provide just about any reference voltage desired.

doesn't come into play, but for Fig. 4.11 it will prove to be central to our analysis. For what follows, I'll take $V_C = 60$ volts.

To start, we'll take C as uncharged, and so initially C has a zero voltage drop across its terminals, and bulbs B1 and B2 are both off (that is, non-conducting). Then, we close S. The bulb node voltages e_1 and e_2 instantly jump to 90 volts but, because B1 and B2 are not perfectly identical, one will strike before the other. Suppose it is B1. Then the (constant) voltage drop across B1 is $V_C = 60$ volts (read the previous paragraph again); e_1 is said to be *fixed* or *clamped* to 60 volts.

Since the voltage drop across C cannot change instantly, then the B2 voltage $e_2 = 60$ volts, too, to keep the instantaneous voltage drop across C equal to zero. Note, carefully, however, that e_2 is *not* clamped (B2 is not conducting) and so can change. With e_1 and e_2 now both less than 90 volts, there is current in both $R1$ and $R2$, with both currents entirely flowing through B1 (remember, B2 is off). The current in $R2$, in particular, flows through C on its way to B1, and so charges C, resulting in e_2 increasing.

Writing Kirchhoff's current law at the e_2 node (not to be an echo chamber, remember that B2 is off and so that bulb's current is zero), and using the fact that $R1 = R2$ in Fig. 4.11 and so I'll call both resistors R, we have

$$\frac{90 - e_2}{R} = C\frac{d}{dt}(e_2 - e_1) = C\frac{de_2}{dt} - C\frac{de_1}{dt}.$$

But since e_1 is *clamped* (which means $\frac{de_1}{dt} = 0$), we then have

$$90 - e_2 = RC\frac{de_2}{dt}. \tag{4.31}$$

Laplace transforming (4.31) gives

$$\frac{90}{s} - E_2 = RC[sE_2 - e_2(0+)]$$

or, as $e_2(0+) = 60$,

$$\frac{90}{s} - E_2 = RCsE_2 - 60RC$$

and so, with a little rearranging,

$$E_2(s) = \frac{90}{s(RCs + 1)} + \frac{60RC}{RCs + 1}. \tag{4.32}$$

Making a partial fraction expansion of the first term on the right-hand-side of (4.32), we have (with A and B as constants to be determined)

$$\frac{90}{s(RCs+1)}=\frac{A}{s}+\frac{B}{RCs+1}.$$

Thus, multiplying through by s,

$$A+\frac{sB}{RCs+1}=\frac{90}{(RCs+1)}$$

and so, setting $s = 0$, we have $A = 90$. Also, multiplying through by $(RCs + 1)$,

$$\frac{A(RCs+1)}{s}+B=\frac{90}{s}$$

and so, setting $s = -1/RC$, we have $B = -90RC$. Thus, (4.32) becomes

$$E_2(s)=\frac{90}{s}-\frac{90RC}{(RCs+1)}+\frac{60RC}{RCs+1}$$

or,

$$E_2(s)=\frac{90}{s}-30\frac{1}{s+\frac{1}{RC}}. \tag{4.33}$$

Returning to the time domain,

$$e_2(t)=90-30e^{-\frac{t}{RC}}, t\geq 0,$$

a result we perhaps should have seen (in retrospect) earlier. In any case, as claimed, e_2 rises from 60 volts (at $t = 0$) towards 90 volts.

The bulb voltage e_2 never gets near 90 volts, however, because when e_2 reaches 75 volts B2 strikes and therefore clamps e_2 to $V_C = 60$ volts. In other words, e_2 instantly drops by 15 volts, and so e_1 must instantly drop by 15 volts as well, because the voltage drop across C cannot change instantly. That is, e_1 goes from 60 volts to 45 volts which, being less than $V_E = 52$ volts, causes B1 to turn off. The situation is now as follows: $e_1 = 45$ volts, B1 is off, e_2 is *clamped* to 60 volts, and B2 is on. Without any loss in generality in what follows, we can think of this situation as defining a new $t = 0$ instant.

Just as before, in our initial calculations, there is current in both $R1$ and $R2$ (both currents flow into B2), with the current in $R1$ in particular charging C which causes e_1 to increase. (Perhaps now you already see what e_1 does, but let's go through the formal steps, just to be sure.) Writing Kirchhoff's current law at the e_1 node (remember that B1 is off and so that bulb's current is zero), we have

$$\frac{90-e_1}{R}=C\frac{d}{dt}(e_1-e_2)=C\frac{de_1}{dt}-C\frac{de_2}{dt}.$$

or, since e_2 is *clamped* (which means $\frac{de_2}{dt}=0$), we then have

$$90 - e_1 = RC\frac{de_1}{dt}, \tag{4.34}$$

or, as $e_1(0+) = 45$ volts, taking the Laplace transform of (4.34) says

$$\frac{90}{s} - E_1 = RC[sE_1 - e_1(0+)] = sRCE_1 - 45RC.$$

Going through the partial fraction expansion as before for E_2, we arrive at

$$e_1(t) = 90 - 45e^{-\frac{t}{RC}}, t \geq 0.$$

So, e_1 rises from 45 volts towards 90 volts. But when it reaches 75 volts B1 strikes and so clamps e_1 to 60 volts. That is, e_1 drops suddenly by 15 volts, and of course so must e_2 (to keep the voltage drop across C instantaneously unchanged). That is, e_2 drops suddenly from 60 volts to 45 volts, and thus B2 turns off. This should all sound pretty familiar by now.

We now observe that *the conditions in the circuit are just as they were at the previous $t = 0$ instant, but with the roles of e_1 and e_2, and of B1 and B2, reversed.*

What happens from now on should be clear: the bulbs B1 and B2 alternately turn on and off at a rate determined by how long it takes the voltage $90 - 45e^{-\frac{t}{RC}}$ (starting at $t = 0$ with a value of 45 volts) to rise to 75 volts. If we call this time T, then

$$90 - 45e^{-\frac{T}{RC}} = 75,$$

an equation easily solved to give

$$T = RCln(3).$$

For the circuit values in Fig. 4.11,

$$T = \left(2.2 \times 10^6\right)\left(10^{-7}\right) \ln(3) = 0.242 \text{ seconds}.$$

The neon bulbs in this circuit (called a *free-running astable multi-vibrator oscillator*) switch back-and-forth from being on to off to on again like two kids on a teeter-totter, spending about a quarter second in one state before transitioning to the other state. The rhythmic, on-off flashes of B1 and B2 are synchronized, but 180° out-of-phase.

4.4 A Constant Current Generator

In this final section of the chapter, we'll combine Heaviside's generalized impedance idea, the Laplace transform, and computer processing, to study the circuit of Fig. 4.12. It appears to be a not overly-complicated one, consisting of just two identical inductors, three identical capacitors, an a-c voltage source, and a single

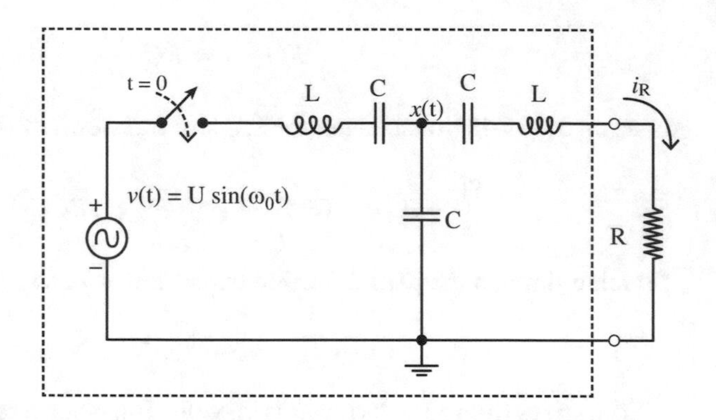

Fig. 4.12 A remarkable circuit

resistor. The frequency of the a-c source is not arbitrary, but rather has been adjusted (you'll see why, soon) so that

$$\omega_0{}^2 = \frac{2}{LC}$$

where, of course,

$$\omega_0 = 2\pi f_0$$

and f_0 is the frequency in hertz. Before the switch is closed at $t = 0$, there is no stored energy in the circuit. To understand why I've enclosed everything but the resistor inside the dashed box, let's calculate the current in the in the resistor: I think you'll be (greatly) surprised at what we find.

If we call the transformed current in the voltage source $I(s)$, then

$$I(s) = \frac{V(s) - X(s)}{sL + \frac{1}{sC}} = [V(s) - X(s)]\frac{sC}{s^2LC + 1}. \tag{4.35}$$

Now, from Kirchhoff's current law at the $x(t)$ node

$$I(s) = \frac{X(s)}{\frac{1}{sC}} + \frac{X(s)}{sL + \frac{1}{sC} + R} = X(s)\left[sC + \frac{sC}{s^2LC + sRC + 1}\right], \tag{4.36}$$

and so, expanding (4.35) and using (4.36),

$$V(s)\frac{sC}{s^2LC + 1} - X(s)\frac{sC}{s^2LC + 1} = X(s)\left[sC + \frac{sC}{s^2LC + sRC + 1}\right]$$

or,

$$V(s)\frac{1}{s^2LC + 1} = X(s)\left[\frac{1}{s^2LC + 1} + 1 + \frac{1}{s^2LC + sRC + 1}\right]$$

or,

$$V(s) = X(s)\left[1 + s^2LC + 1 + \frac{s^2LC + 1}{s^2LC + sRC + 1}\right]$$

which becomes, after some algebra,

$$V(s) = X(s)\left[\frac{s^4(LC)^2 + s^3RLC^2 + s^24LC + s2RC + 3}{s^2LC + sRC + 1}\right]. \tag{4.37}$$

Solving for $X(s)$,

$$X(s) = V(s)\frac{s^2LC + sRC + 1}{s^4(LC)^2 + s^3RLC^2 + s^24LC + s2RC + 3}. \tag{4.38}$$

The transformed current in the resistor is

$$I_R(s) = \frac{X(s)}{sL + \frac{1}{sC} + R} = X(s)\frac{sC}{s^2LC + sRC + 1}$$

or, using (4.38),

$$I_R(s) = V(s)\frac{sC}{s^4(LC)^2 + s^3RLC^2 + s^24LC + s2RC + 3}$$

where, from the transform pair in (3.16),

$$V(s) = U\frac{\omega_0}{s^2 + \omega_0{}^2}.$$

Thus, looking back at (4.38), we have

$$I_R(s) = U\omega_0C\frac{s}{(s^2 + \omega_0{}^2)\left[s^4(LC)^2 + s^3RLC^2 + s^24LC + s2RC + 3\right]}. \tag{4.39}$$

The next, traditional step in this analysis would be to invert $I_R(s)$ back to $i_R(t)$ by doing a partial fraction expansion, a step that looks pretty intimidating because the denominator in (4.39) is a *sixth* degree polynomial! Of course it is already partially factored with that $(s^2 + \omega_0{}^2)$ factor, but we are still left with the fourth degree polynomial in the square brackets. Now, before you fall over in a dead faint, let me tell you we are *not* going to factor that fourth degree polynomial. Since all the coefficients are symbolic (except for the 3), that is, the coefficients are in terms of R, L, and C, such a factoring would be pretty hard to do. But even if we had numerical values for the coefficients, factoring a fourth degree polynomial isn't a trivial task unless we have a computer handy (look back at note 1).[7]

You, of course, know what we are going to do instead — we are going to turn our standard inverse-Laplace transform *MATLAB* computer code loose on (4.39). Before I do that, however, let me outline how a traditional analysis would proceed, and that will make you appreciate the computer code approach even more. Our first step would be to write

$$I_R(s) = U\omega_0 C\left[\frac{As+B}{s^2+\omega_0{}^2} + \frac{Ds^3+Es^2+Fs+G}{s^4(LC)^2+s^3RLC^2+s^24LC+s2RC+3}\right], \tag{4.40}$$

where A, B, D, E, F, and G are constants to be determined. One way to do that is to put (4.40) over a single denominator (by cross-multiplying) and then setting the coefficient of each power of s in the resulting numerator equal to the coefficient of the corresponding power of s in the numerator of (4.39). This gives six equations in six unknowns:

$$\begin{aligned}
s^5 &: \; A(LC)^2 + D = 0\\
s^4 &: \; A\left(RLC^2\right) + B(LC)^2 + E = 0\\
s^3 &: \; A(4LC) + B\left(RLC^2\right) + F + D\omega_0{}^2 = 0\\
s^2 &: \; A(2RC) + B(4LC) + G + E\omega_0{}^2 = 0\\
s^1 &: \; A(3) + B(2RC) + F\omega_0{}^2 = 1\\
s^0 &: \; B(3) + G\omega_0{}^2 = 0.
\end{aligned}$$

This system is, of course, theoretically solvable by Cramer's rule since the number of equations equals the number of unknowns but, holy cow, what a job! To solve for each variable requires doing a separate six-by-six determinant for a numerator, divided by a common denominator that also requires doing a six-by-six determinant (the so-called *system determinant*). If you are lucky enough to stumble on the idea of solving first for G, you'll find that $G = 0$ *if* $\omega_0{}^2 = \frac{2}{LC}$ (and so here is where that seemingly out-of-left-field frequency makes its first appearance), which in turn implies that $B = 0$, which then (after yet more algebra) lets you determine that $A = -1$. (I'll let you confirm all these claims, if you are so inclined.) But that still leaves D, E, and F. What are *they*?

One possible answer is — maybe we don't care. That no doubt must seem like heresy in a book on transients, *unless* you take the position that it is the *non*-transient behavior of the circuit that is of real interest, and that all we really want to know about the transient behavior is: how long does it last? That entire second term in the

---

[7] A standard feature in all 'early' (that is, pre-*MATLAB*) electrical engineering textbooks was an appendix on how to factor polynomials of degree greater than two. Those appendices were (in my opinion) the ultimate in brain-deadening drudgery to read.

square brackets of (4.40), where D, E, F, and G appear, might be nothing but transient terms which will eventually disappear.

To determine if that entire second term *is* entirely transient, we don't actually have to know D, E, F and G. All we have to know is that each of the factors of the fourth degree denominator polynomial has a negative real part. That will insure, if we do a partial fraction expansion of the second term, that the time domain term associated with each factor will exponentially decay.[8] And, as a demonstration that the number gods are not completely malevolent, there *is* a wonderful algorithm that determines if a polynomial of *any* degree has all its factors with negative real parts, through an examination of nothing but the coefficients of the polynomial. Called the *Routh-Hurwitz algorithm*,[9] it involves only the easy calculation of a series of two-by-two determinants.

Okay, with 'what we would do if we didn't have a computer' out of the way, here's what a computer can do. The *MATLAB* code **constant.m** calculates and plots $i_R(t)$ from (4.39), assuming that the peak amplitude of the voltage source is $U = 1$ volt, $L = 1$ mH $= 1,000\ \mu$H, $C = 0.5\ \mu$F (these values for L and C give a value of $f_0 = 10.066$ kHz), and $R = 20$ ohms. (All these values have been arbitrarily picked.) The resistor current is shown, for the first 600 microseconds, in Fig. 4.13 and you see that, after an initial transient of about 300 microseconds duration, the current settles down to a steady-state of a pure sinusoid at frequency f_0 with a peak amplitude of about 0.03 $A = 30\ mA$.

```
%constant.m/created by PJNahin for Electrical Transients
(8/15/2017)
R=20;L=1000;C=0.5;U=1;
w02=2/(L*C);k0=U*C*sqrt(w02);
k1=(L*C)^2;k2=R*L*(C^2);k3=4*L*C;k4=2*R*C;
syms s t
h=ilaplace(k0*s/((s^2+w02)*(k1*s^4+k2*s^3+k3*s^2+k4*s
+3)));
t=linspace(0,600,250);
current=subs(h);
current=vpa(current);
plot(t,current,'-k')
xlabel('time in microseconds')
ylabel('resistor current in amperes')
```

[8]Just to be clear, a factor is of the form $(s - \alpha)$, where the *real part* referred to is the real part of α. Any factor with a zero real part will be one of a *pair* of imaginary conjugates that, together, will give an *undamped* sinusoid. Of course, there will be no factors with a positive real part because that would mean $i_R(t)$ would have a term that *grows* with time and, for a circuit made only of passive components, that is physically impossible.

[9]Named after the English mathematician Edward Routh (1831–1907) who formulated the algorithm in 1887, and the German mathematician Adolf Hurwitz (1859–1919) who independently made the same discovery in 1895. You can find their algorithm in any good book on control theory (there are several people who have put free-to-use computer codes on the Web that implement the algorithm).

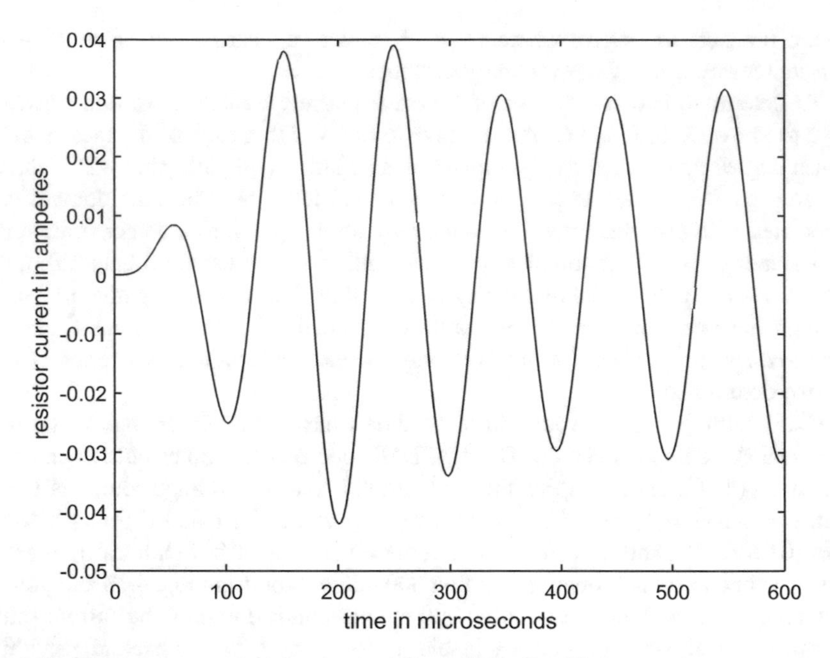

Fig. 4.13 The resistor current in the circuit of Fig. 4.12 for $R = 20$ ohms

What makes this *really* remarkable, however, is what the code produces for other values of R. Figure 4.14 shows $i_R(t)$ for $R = 40$ ohms, 100 ohms, 200 ohms, and 300 ohms. In each case the current settles down, after an initial transitory interval (which increases in duration as R increases) to the *same peak amplitude sinusoid*! We appear to have a constant a-c current generator (same output current, independent of R), and so now you see why the code has the name it does. This is such 'interesting' behavior, from such a simple circuit, that it is worth a little time to see if we can understand it from fundamental, basic, a-c steady-state concepts. For that, I'll assume you recall from previous studies that the a-c steady-state impedances for an inductor L and a capacitor C are $i\omega L$ and $\frac{1}{i\omega C}$, respectively, where of course $i = \sqrt{-1}$.

Figure 4.15 shows the circuit of Fig. 4.12, with the capital letters U, X, I_s, and I_R representing a-c steady-state peak values. That is, they are complex-valued quantities that electrical engineers call *phasors* (for *phase vectors*). The voltage source current is

$$I_s = \frac{U - X}{i\omega L + \frac{1}{i\omega C}} = \frac{i\omega C}{1 - \omega^2 LC}(U - X). \tag{4.41}$$

That current flows into the node with voltage X, where it splits along two paths to ground (with one of those paths being that of the current in R). In such a case the fraction of the current that flows into *one* path is the ratio of the impedance of the *other* path to the sum of the two path impedances. So,

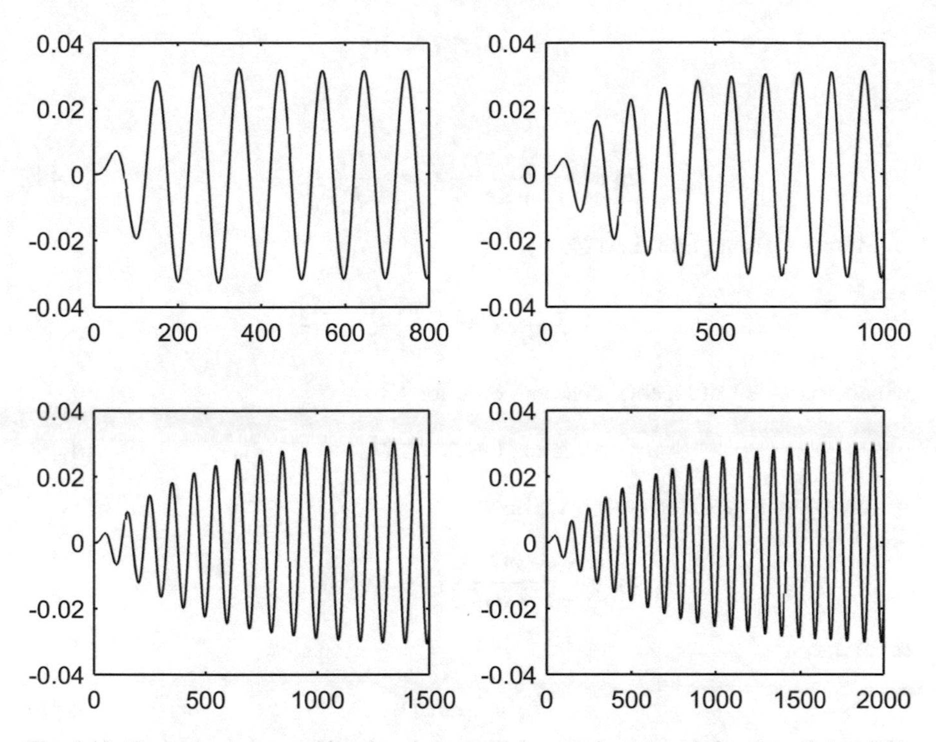

Fig. 4.14 The resistor current (time in microseconds/current in amperes) for $R = 40$ (top left), $R = 100$ (top right), $R = 200$ (bottom left), and $R = 300$ ohms (bottom right)

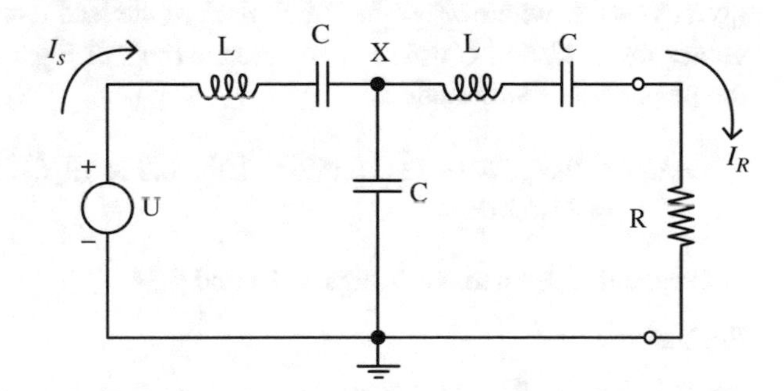

Fig. 4.15 Figure 4.12 in the a-c steady-state

$$I_R = \frac{X}{i\omega L + \frac{1}{i\omega C} + R} = I_s \frac{\frac{1}{i\omega C}}{i\omega L + \frac{1}{i\omega C} + R + \frac{1}{i\omega C}} = \frac{1}{2 - \omega^2 LC + i\omega RC} I_s.$$

That is,

$$I_R = \frac{i\omega C}{1 - \omega^2 LC + i\omega RC} X = \frac{1}{2 - \omega^2 LC + i\omega RC} I_s. \tag{4.42}$$

Now, suppose we set $2 - \omega^2 LC = 0$. Then, (4.41) and (4.42) become

$$I_s = -i\omega C(U - X) \tag{4.43}$$

and

$$I_R = \frac{i\omega C}{-1 + i\omega RC} X = \frac{1}{i\omega RC} I_s. \tag{4.44}$$

Putting I_s from (4.43) into (4.44),

$$\frac{i\omega C}{-1 + i\omega RC} X = \frac{-i\omega C(U - X)}{i\omega RC}$$

which, with a bit of algebra, can be solved for X to give

$$X = (1 - i\omega RC)U. \tag{4.45}$$

Then, using (4.45) in (4.44), we have

$$I_R = \frac{i\omega C}{-1 + i\omega RC}(1 - \mathrm{i\omega RC})U$$

or, at last,

$$I_R = -i\omega CU. \tag{4.46}$$

That is, I_R does not depend on R and we do indeed have a constant current generator with the peak amplitude of the steady-state sinusoidal current[10] being given by ωCU where $\omega^2 = \frac{2}{LC}$. (Look back at the lead factor in (4.39).) Inserting the values for ω, C, and U used to generate the plots of Figs. 4.13 and 4.14, we see that the peak current amplitude is

$$\begin{aligned}\omega CU &= 2\pi f_0 CU = 2\pi\left(10.066 \times 10^3\right)\left(0.5 \times 10^{-6}\right)(1) = 0.0316 \text{ amperes} \\ &= 31.6 \text{ mA},\end{aligned}$$

in excellent agreement with Figs. 4.13 and 4.14.

Problems

4.1 The energy dissipated in the load resistor R of the constant current generator of Fig. 4.15 must (since we believe energy is conserved) come from the voltage source. Find an expression for the power output of the voltage source and show it is equal to the power dissipated by R. (Of course, if the voltage source is not capable of providing the required power, the circuit is no longer a constant current generator.)

[10]The physical significance of the $-i$ factor in (4.46) is that there is a 90° phase difference between the source voltage and the load current.

Chapter 5
Transmission Lines

5.1 The Partial Differential Equations of Transmission Lines

So far, all of our work has been with circuits made from what are called *lumped* components. That is, from inductors, capacitors, and resistors that can be thought of as each having their behavior concentrated at a point in space. The time domain analysis of such circuits involves the mathematics of *ordinary* differential equations, in which there is but a single independent variable, t for time. In this final chapter of this introductory book I'll show you the very elementary beginnings of the theory of circuits constructed from *distributed* components; that is, circuits in which inductance, capacitance, and resistance are *not* located at discrete points, but rather are *smeared out* in space along one (or more) spatial directions. *Now* when we talk of a current or a voltage in such a circuit, we won't just specify its value as a function of time, but also as a function of space, as well. That is, as $i(x, t)$ or $v(x, t)$. The mathematics of such circuits is that of *partial* differential equations, because there are multiple (for us, just two, x and t) independent variables.

In this chapter we'll only just dip our toes into a discussion of transmission lines, as the subject is so broad as to require a substantial book of its own. Indeed, back in my student days at Stanford it was an entirely separate course: the very next term after EE116 on transients (the experience that inspired this book), I took EE117 on transmission lines. In this chapter, then, I'll tell you just enough for you to gain an appreciation for the richness of the subject, and for you to be able to follow the historically important transient analyses of the Atlantic Cable, and of Heaviside's discovery of the infinitely long distortionless transmission line. We'll end with a brief look at how adding the realism of a finite length complicates matters.

The geometry of a two-conductor (this includes coaxial) transmission lines is given in Fig. 5.1, which shows an arbitrary, very short (Δx) length of the line. The so-called *distributed parameters* of the line are

P. J. Nahin, *Transients for Electrical Engineers*,
https://doi.org/10.1007/978-3-319-77598-2_5

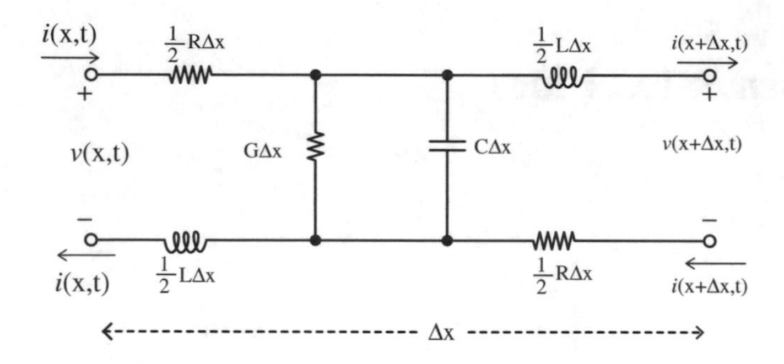

Fig. 5.1 Geometry of a two-conductor transmission line

R = resistance per unit length (ohmic resistance of the conductors)
L = inductance per unit length (self *and* mutual inductance of the conductors)
G = conductance per unit length (leakage path resistance between the conductors)
C = capacitance per unit length (capacitive coupling between the conductors)

where length is measured along the *two* conductors (which is why the resistance and inductance in *each* wire is $\frac{1}{2}R\Delta x$ and $\frac{1}{2}L\Delta x$, respectively, in Fig. 5.1), while the capacitance and conductance[1] are shunt (parallel) effects *between* the conductors.

Starting at the upper-left point of Fig. 5.1 (where the current $i(x,t)$ enters the upper-left $\frac{1}{2}R\Delta x$ resistance), and summing voltage drops around the outer closed loop of the circuit, Kirchhoff's voltage law says

$$i(x,t)\frac{1}{2}R\Delta x+\frac{1}{2}L\Delta x\frac{\partial i(x+\Delta x,t)}{\partial t}+v(x+\Delta x,t)+i(x+\Delta x,t)\frac{1}{2}R\Delta x \\ +\frac{1}{2}L\Delta x\frac{\partial i(x,t)}{\partial t}-v(x,t)=0, \tag{5.1}$$

where *partial* derivatives are used because we have two independent variables. With a bit of rearranging, (5.1) becomes

$$-[v(x+\Delta x,t)-v(x,t)]=\frac{1}{2}R\Delta x[i(x,t)+i(x+\Delta x,t)] \\ +\frac{1}{2}L\Delta x\frac{\partial}{\partial t}[i(x+\Delta x,t)+i(x,t)]. \tag{5.2}$$

Then, dividing through (5.2) by Δx, we have

[1]Conductance is the reciprocal of resistance, and at one time had the units of *mhos* (ohms spelled backward, thus showing that electrical engineers are not entirely without a sense of humor). In 1971, however, an international committee decided that was a bit too flippant and we now measure conductance in the units of *siemens*, named after the German industrialist Ernst Werner Siemens (1816–1892). (Hard-core electrical engineers still use mhos, however.)

$$-\frac{[v(x+\Delta x,t)-v(x,t)]}{\Delta x}=\frac{1}{2}R[i(x,t)+i(x+\Delta x,t)]+\frac{1}{2}L\frac{\partial}{\partial t}[i(x+\Delta x,t)+i(x,t)]$$

and so, if we let $\Delta x \rightarrow 0$, we have $i(x + \Delta x, t) \rightarrow i(x, t)$ and we arrive at

$$-\frac{\partial v}{\partial x}=iR+L\frac{\partial i}{\partial t}. \tag{5.3}$$

Once we have (5.3) down on paper we realize (perhaps) that we should have been able to skip the math and to have argued directly that the *loss* of voltage (the meaning of the minus sign) along the line per unit length is equal to the ohmic voltage drop per unit length plus the inductive voltage drop per unit length. With that physical insight as inspiration, we can immediately write a similar equation for the *loss* of current along the line per unit length as equal to the leakage current between the two conductors per unit length plus the current needed to charge the line's capacitance between the two conductors per unit length. That is,

$$-\frac{\partial i}{\partial x}=Gv+C\frac{\partial v}{\partial t}. \tag{5.4}$$

If we differentiate (partially) our two equations, (5.3) with respect to x, and (5.4) with respect to t, we have

$$-\frac{\partial^2 v}{\partial x^2}=R\frac{\partial i}{\partial x}+L\frac{\partial^2 i}{\partial x\partial t} \tag{5.5}$$

and

$$-\frac{\partial^2 i}{\partial t\partial x}=G\frac{\partial v}{\partial t}+C\frac{\partial^2 v}{\partial t^2}. \tag{5.6}$$

Since x and t are independent, we can safely assume that the order of differentiation doesn't matter; that is,

$$\frac{\partial^2 i}{\partial x\partial t}=\frac{\partial^2 i}{\partial t\partial x}.$$

Thus, putting (5.4) and (5.6) into (5.5),

$$-\frac{\partial^2 v}{\partial x^2}=R\left[-Gv-C\frac{\partial v}{\partial t}\right]+L\left[-G\frac{\partial v}{\partial t}-C\frac{\partial^2 v}{\partial t^2}\right]$$

which is a second-order partial differential equation called the *first telegraphy equation* (because of its obvious connection to the behavior of the wire communication links at the heart of nineteenth century telegraph land lines):

$$\frac{\partial^2 v}{\partial x^2} = LC\frac{\partial^2 v}{\partial t^2} + (RC + LG)\frac{\partial v}{\partial t} + RGv. \tag{5.7}$$

In much the same way, you can alternatively eliminate v to get a second-order partial differential equation for i, known as the *second telegraphy equation*:

$$\frac{\partial^2 i}{\partial x^2} = LC\frac{\partial^2 i}{\partial t^2} + (RC + LG)\frac{\partial i}{\partial t} + RGi. \tag{5.8}$$

The structural simularity of the two telegraphy equations is striking.

There are several special cases of the telegraphy equations that are of particular interest. One, called the *lossless line*, assumes $R = G = 0$. In that case, (5.7) reduces to

$$\frac{\partial^2 v}{\partial x^2} = LC\frac{\partial^2 v}{\partial t^2} = \frac{1}{c^2}\frac{\partial^2 v}{\partial t^2} \tag{5.9}$$

where

$$c = \frac{1}{\sqrt{LC}}$$

has the units of *speed*[2]: (5.9) is called the *wave equation*, as it occurs in the theory of electromagnetic waves where, if the 'transmission line' is actually empty space, c is the speed of light. We will *not* discuss this case in this book.

A special case we *will* discuss, a lot, is the historically important one of the Trans-Atlantic telegraph cable, which is technically an example of the so-called leakage-free ($G = 0$), non-inductive ($L = 0$) case. Under those assumptions (5.7) reduces to

$$\frac{\partial^2 v}{\partial x^2} = RC\frac{\partial v}{\partial t}, \tag{5.10}$$

which is called the *diffusion equation* because it describes how various physical quantities diffuse or spread in absorbing mediums (ink in water, or heat in metal, for example). The diffusion equation was known to mathematical physicists *long* before the first transmission line was constructed, and the theoretical methods they developed before 1800 to solve it were of immense aid to the electrical engineers who later created the Atlantic Cable.

At the 'slow' signalling speeds of a human-operated telegraph key (a dozen or so words per minute), inductive effects were negligible in the Atlantic Cable, and conductance was pretty small, too (that is, the leakage resistance between the conductors was very large). Thus, only R and C were considered to be the important parameters for the Atlantic Cable and so we have (5.10). That cable, about 2,000

[2]Can you show this? Hint: look at the dimensions on each side of (5.9).

nautical miles in length, spanned the Atlantic Ocean to connect the land lines of Ireland in the Old World with those in Newfoundland in the New World. The Cable was rightfully considered to be one of the greatest achievements of nineteenth century engineering and science, and understanding how it worked depends on being able to solve (5.7) and (5.8).

So, how *do* we mathematically handle (5.7) and (5.8)? First of all, the fact is that despite all the work we've put into the Laplace transform, the very first transient analysis of a transmission line was done in the time domain. That pioneering analysis, done by William Thomson (look back at Sect. 1.4) in 1854 and published in 1855, was the original theoretical basis for submarine transmission cable lines.[3] Thomson based his analysis on the work in thermodynamics done by the French mathematical physicist Joseph Fourier. Thomson found that Fourier's so-called *heat equation* (the diffusion equation) applies just as well (under certain conditions that were more or less satisfied by the Atlantic Cable) to electricity[4]: he found mathematical expressions in Fourier's book for a number of situations that could be directly translated into transmission line problems, and he took great advantage of those existing solutions. The pioneers who found those solutions used classical, time domain methods (see Appendix 3 for how to do that for the case of a unit step voltage input), but here we'll do it the modern way, with the Laplace transform, in our analyses of the transient behavior of transmission lines in response to the closing of a telegraph key.

5.2 Solving the Telegraphy Equations

We start by Laplace transforming (5.7), the partial differential equation for the voltage along a *general* line (all four line parameters present) where, using various transform results from Chap. 3,

$$\mathcal{L}\{v(x,t)\} = \int_0^\infty v(x,t)e^{-st}dt = V(x,s),$$

[3]W. Thomson, "On the Theory of the Electric Telegraph," *Proceedings of the Royal Society of London*, May 1855, pp. 382–399. For his contributions to the Atlantic Cable project, Thomson was knighted by Queen Victoria in 1866 (becoming Sir William), and later (1892) was elevated to the peerage to become Baron Kelvin. Lord Kelvin is one of the giants in the world of physics and, at the end of his life, he was buried in Westminster Abby (an honor reserved by England for her greatest heroes), to lie forever near the supremely great Isaac Newton.

[4]Fourier's 1822 book *Analytical Theory of Heat* had an enormous influence on Thomson, who had read it when still a teenager (1839): "In a fortnight I had mastered it — gone right through it" and he rightfully declared it to be a "mathematical poem."

$$\mathcal{L}\left\{\frac{\partial v}{\partial t}\right\} = sV(x,s) - v(x,0+),$$

$$\mathcal{L}\left\{\frac{\partial^2 v}{\partial t^2}\right\} = s^2V(x,s) - sv(x,0+) - \frac{dv(x,t)}{dt}\mid_{t=0+},$$

$$\mathcal{L}\left\{\frac{\partial^2 v}{\partial x^2}\right\} = \frac{\partial^2}{\partial x^2}[\mathcal{L}\{v(x,t)\}] = \frac{\partial^2}{\partial x^2}V(x,s) = \frac{d^2}{dx^2}V(x,s).$$

The difference between the last two lines — note which variable the derivatives are with respect to — should be carefully understood. Thus, (5.7) becomes

$$\frac{d^2V(x,s)}{dx^2} = LC\left[s^2V(x,s) - sv(x,0+) - \frac{dv(x,t)}{dt}\mid_{t=0+}\right]$$
$$+(RC+LG)[sV(x,s) - v(x,0+)] + RGV(x,s)$$

or, after a bit of rearranging,

$$\frac{d^2V(x,s)}{dx^2} - (Ls+R)(Cs+G)V(x,s) = -L\left[Gv(x,0+) + C\frac{dv(x,t)}{dt}\mid_{t=0+}\right] \tag{5.11}$$
$$-C[Ls+R]v(x,0+).$$

At this point, there is a subtle technical issue that needs to be addressed before continuing. The value of $\frac{dv(x,t)}{dt}\mid_{t=0+}$, the initial rate of change of the voltage with respect to time, at an arbitrary location along the line, is usually *not* obvious. We can side-step this problem if we work instead with the initial rate of change with respect to x of the *current*, that is, with $\frac{di(x,t)}{dx}\mid_{t=0+}$ — I'll elaborate on this claim in just a moment — and we can get that value from (5.4). So, evaluating (5.4) at $t = 0+$ (note, *carefully*, that we are simply substituting $t = 0+$ and are *not* Laplace transforming), we have

$$-\frac{di(x,t)}{dx}\mid_{t=0+} = Gv(x,0+) + C\frac{dv(x,t)}{dt}\mid_{t=0+}$$

and so

$$C\frac{dv(x,t)}{dt}\mid_{t=0+} = -\frac{di(x,t)}{dx}\mid_{t=0+} - Gv(x,0+).$$

Putting this into (5.11) we arrive at

$$\frac{d^2V(x,s)}{dx^2} - \gamma^2V(x,s) = L\frac{di(x,t)}{dx}\mid_{t=0+} - C[Ls+R]v(x,0+) \tag{5.12}$$

where

$$\gamma = \sqrt{(Ls + R)(Cs + G)}.$$

The reason (5.12) is preferable to (5.11) is that the initial value of $\frac{di(x,t)}{dx}$ is, as I claimed earlier, usually more easily arrived at than is the initial value of $\frac{dv(x,t)}{dt}$. To see this, consider the problem of calculating $v(x, t)$ on a transmission line (in response to a given input signal $v(0, t)$), a line that we are told is 'initially relaxed' (that is, there is no current or stored charge in the line at $t = 0$). We can immediately say, for such a line, that

$$v(x, 0+) = 0, x > 0$$

and

$$i(x, 0+) = 0, x > 0.$$

As time increases from $t = 0+$, we obviously expect the cable current at a given location to change from zero. But suppose we keep time constant, and look at the cable current as we change x. As long as $x > 0$ (that is, we are not at the beginning of the cable) then as we move along the cable with increasing x we will continue to see $i = 0$ until some time $t > 0+$. Thus, we can definitely say

$$\frac{di(x,t)}{dx} \Big|_{t=0+} = 0, x > 0.$$

Next, consider the case of the cable voltage $v(x, t)$. As time increases from $t = 0+$ we again obviously expect the cable voltage at a given point to change from zero. Suppose now that we keep x constant (we observe a fixed point on the cable) and ask how the voltage there changes *with respect to time*? That is, what is $\frac{dv(x,t)}{dt} \big|_{t=0+}$? Alas, knowing that $v(x, 0+) = 0$ for any $x > 0$ doesn't tell us anything about how the voltage changes *with time*. It isn't the distinction between voltage and current that is important in this consideration, but rather which variable we are differentiating with respect to (x or t). Differentiating with respect to x, and with respect to t, makes a *big* difference.

To complete this section, we'll calculate $v(x, t)$ and $i(x, t)$ for an initially relaxed, infinitely long[5] transmission line when a unit step voltage is applied to the $x = 0$ end at $t = 0$. The general problem (for arbitrary values of R, L, C, and G) was first attempted by Heaviside, using his operational calculus, but his solution was later questioned (after his death) by others.[6] We'll work our way towards the general

[5]To account for a finite length is not conceptually difficult, but it does add considerably to the mathematics. I'll say more on this at the end of the chapter.

[6]See, for example, F. W. Carter, "Note on Surges of Voltage and Current in Transmission Lines," *Proceedings of the Royal Society*, Series A (volume 156), August 1936, pp. 1–5.

solution using the Laplace transform, with our first discussion dealing with the Atlantic Cable, then with Heaviside's distortionless line, and finally we'll move on to end with the general cable.

From (5.12), then, we start with

$$\frac{d^2V(x,s)}{dx^2} - \gamma^2 V(x,s) = 0, v(0,t) = u(t). \tag{5.13}$$

This is an ordinary differential equation that is easily solved. Specifically, assume a solution of the form

$$V(x,s) = Ke^{px}$$

where K and p are some constants. Then

$$Kp^2e^{px} - \gamma^2 Ke^{px} = 0$$

and so

$$p = \pm\gamma.$$

That is,

$$V(x,s) = K_1e^{\gamma x} + K_2e^{-\gamma x} \tag{5.14}$$

where

$$\gamma = \sqrt{(Ls+R)(Cs+G)}.$$

For an *infinite* line we demand that $K_1 = 0$ as otherwise $V(x, s)$ would become unbounded as $x \to \infty$. So,

$$V(x,s) = K_2e^{-\gamma x}. \tag{5.15}$$

Since at $x = 0$ we have $v(0, t) = u(t)$, then

$$V(0,s) = \mathcal{L}\{u(t)\} = \frac{1}{s} = K_2$$

and so

$$V(x,s) = \frac{1}{s}e^{-\gamma x}. \tag{5.16}$$

5.3 The Atlantic Cable

In this case we have $L = G = 0$, which means

$$\gamma = \sqrt{RCs},$$

and so (5.16) becomes

$$V(x,s) = \frac{1}{s}e^{-x\sqrt{RCs}}.$$

Looking back at (3.79), we have the pair

$$1 - erf\left(\frac{a}{2\sqrt{t}}\right) \leftrightarrow \frac{1}{s}e^{-a\sqrt{s}}$$

and so, with

$$a = x\sqrt{RC}$$

we see that

$$v(x,t) = 1 - erf\left(\frac{x\sqrt{RC}}{2\sqrt{t}}\right).$$

That is, the unit step voltage response is

$$v(x,t) = 1 - erf\left(\frac{x}{2}\sqrt{\frac{RC}{t}}\right) \text{ volts.} \tag{5.17}$$

To find the current $i(x, t)$ on the Atlantic Cable, look back at (5.3), set $L = 0$, and then Laplace transform to get

$$-\frac{dV(x,s)}{dx} = I(x,s)R$$

and so

$$I(x,s) = -\frac{1}{R}\frac{dV(x,s)}{dx} = -\frac{1}{R}\left(\frac{-\sqrt{RCs}\,e^{-x\sqrt{RCs}}}{s}\right)$$

or,

$$I(x,s) = \sqrt{\frac{C}{Rs}}\,e^{-x\sqrt{RCs}}. \tag{5.18}$$

To get $i(x, t)$ from (5.18), recall two of our results from earlier: (3.8) and (3.85). That is, repeating them here,

$$tf(t) \leftrightarrow -\frac{d}{ds}F(s) \tag{5.19}$$

and

$$\frac{ae^{-a^2/4t}}{2\sqrt{\pi t^3}} \leftrightarrow e^{-a\sqrt{s}}. \tag{5.20}$$

This last pair is equivalent to

$$\frac{e^{-a^2/4t}}{\sqrt{\pi t^3}} \leftrightarrow \frac{2}{a}e^{-a\sqrt{s}}$$

and so, from (5.19),

$$t\frac{e^{-a^2/4t}}{\sqrt{\pi t^3}} \leftrightarrow -\frac{2}{a}\left(-a\frac{1}{2}s^{-1/2}\right)e^{-a\sqrt{s}}$$

or,

$$\frac{e^{-a^2/4t}}{\sqrt{\pi t}} \leftrightarrow \frac{1}{\sqrt{s}}e^{-a\sqrt{s}}. \tag{5.21}$$

Thus, once again setting $a = x\sqrt{RC}$, we immediately have

$$i(x,t) = \sqrt{\frac{C}{\pi Rt}}e^{-x^2RC/4t} \text{ amperes.} \tag{5.22}$$

Figure 5.2 shows the current $i(x, t)$ on the Atlantic Cable at $x = 2{,}000$ nautical miles (nm). The values of $R = 3$ ohms/nm and $C = 0.5$ μF/nm are 'in the ball park' for that transmission line,[7] and the figure shows the current for the first ten seconds after the start of the unit step input voltage at $x = 0$. There are two features in Fig. 5.2 of practical engineering importance.

First, there is a significant delay of three seconds in the occurrence of the peak current. That is, closing a telegraph key at time $t = 0$ doesn't result in a maximum current response at $x = 2{,}000$ nm until three seconds later. Since the receiver at the $x = 2{,}000$ nautical mile end of the cable, whatever its particular technology may be,

[7] In his famous book *The Theory of Sound* (1894), the great English mathematical physicist John William Strutt (1842–1919), better known in the world of physics as Lord Rayleigh, stated that the RC product on the Atlantic Cable was about $5 \times x10^{-17}$ in cgs (centimeter-gram-second) units. The values for R and C that I give in the text are, in fact, in agreement with Lord Rayleigh's value. Can you show this (see Problem 5.1)?

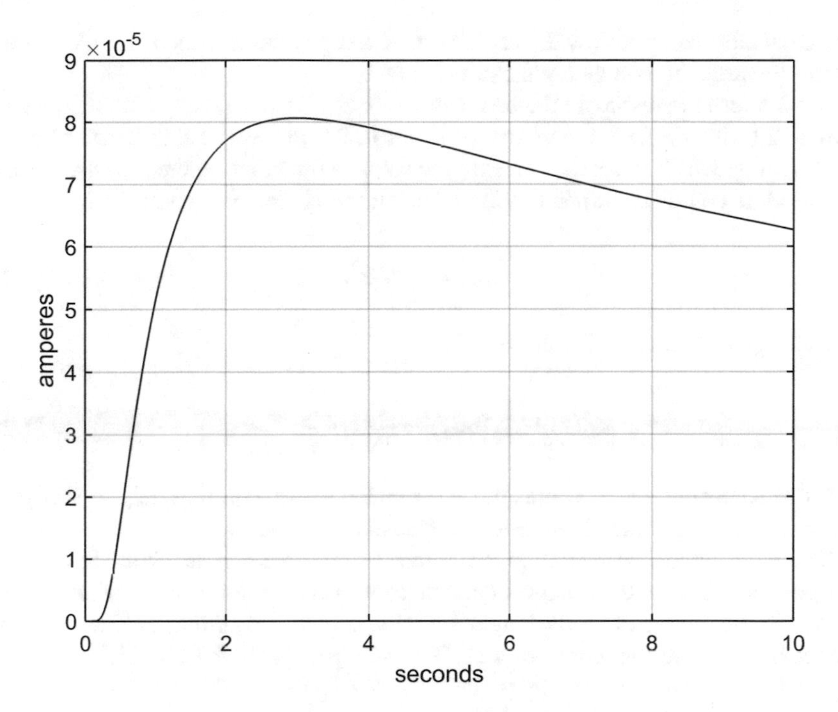

Fig. 5.2 The current at $x = 2,000$ nm on the Atlantic Telegraph Cable. Note the time delay until the current is maximum in response to a unit step voltage input at $t = 0$

will respond to the received *current* (more specifically, to the magnetic field of the current), this delay introduces an upper limit on the speed of transmitting information over the cable. Second, the maximum current is a *substantial* current: Figure 5.2 shows that a one-volt step input produces a maximum current that is slightly in excess of 80 μA, and so if the actual sending voltage would be, say, 40 volts (typical of the voltages actually used on the Atlantic Cable), there would be a maximum current, 2, 000 nm distant, three seconds later, of more than three milliamperes, an easily detectable current in the mid-nineteenth century.[8]

We can understand the general behavior of $v(x, t)$ and $i(x, t)$ by simply looking at the mathematical behavior of (5.17) and (5.22). For any value of x, at $t = 0$, we see from (5.17) that $v(x, 0) = 0$ and that, as t increases, so does $v(x, t)$ because the error function monotonically increases with increasing argument. Since $erf(\infty) = 1$, then $v(x, t) \to 1$ as $t \to \infty$, for any $x > 0$. That is, as time increases the cable voltage

[8]The detection device used on the Atlantic Cable was the *marine* (or *mirror*) *galvanometer* (essentially a light beam bounced off a mirror attached to a tiny magnet hanging by a silk thread inside a coil of wire carrying the received current). The interaction of the magnet's field with the current produced a torque that rotated the magnet (and its mirror). This clever gadget was the invention of the always ingenious William Thomson.

monotonically increases (as the cable current charges the cable's distributed capacitance) towards the unit step voltage value of 1.

Much more interesting is the behavior of $i(x, t)$. For any given value of x, we see from (5.22) that $i(x, 0) = 0$ *and* that $i(x, \infty) = 0$. So, for every x there must be a time $t = t_{max}$ at which $i(x, t)$ reaches an extreme value, a time that depends on the value of x. Indeed, if you differentiate (5.22) and set the result to zero, you'll find that

$$t_{max} = \frac{1}{2}RCx^2. \tag{5.23}$$

Putting (5.23) into (5.22) gives us

$$i_{max}(t_{max}) = \frac{1}{Rx}\sqrt{\frac{2}{\pi e}} \tag{5.24}$$

and, if you substitute $R = 3$ ohms/nm and $x = 2,000$ nm, you'll get $i_{max} = 80.7\ \mu A$ at $t_{max} = 3$ seconds, values in excellent agreement with Fig. 5.2.

The fact that t_{max} depends on *x squared* became known as Thomson's *law of squares*, and it was the source of much controversy among those who invested money in the laying of really long submarine cables. According to Thomson (see Fig. 5.3), doubling the length of a cable would *quadruple* the time delay (thereby

Fig. 5.3 William Thomson, about March 1859, in the early years of his involvement with the Atlantic Cable project. A note (dated November 23, 1892) on the photo in his own hand says the papers he is reading concerned experiments on submarine cables. (Photo courtesy of the Institution of Electrical Engineers (London))

limiting the rate at which a commercial cable could produce revenue). In his 1855 paper (note 3), Thomson mentions that a 200 mile long submarine cable between Greenwich (England) and Brussels (Belgium) had a $\frac{1}{10}$*th* second delay, and then used his law of squares to predict that "the retardation in a cable of [equal construction], extending half round the world (14,000 miles), would be $\left(\frac{14000}{200}\right)^2 \times \frac{1}{10} = 490$ seconds, or $8\frac{1}{6}$ minutes." That's a long time to send *one* Morse code character![9]

5.4 The Distortionless Transmission Line

Suppose we have an infinitely long, initially relaxed cable with distributed parameters that satisfy Heaviside's condition from Appendix 2. If we apply the *arbitrary* voltage $v(0, t)$ to the sending end at $x = 0$, our problem is the determination of the current $i(x, t)$ at an *arbitrary*, distant point. We have

$$\gamma = \sqrt{(Ls + R)(Cs + G)} = (Ls + R)\sqrt{\frac{Cs + G}{Ls + R}} = (Ls + R)\sqrt{\frac{C\left(s + \frac{G}{C}\right)}{L\left(s + \frac{R}{L}\right)}}$$

or, using Heaviside's condition,

$$\gamma = (Ls + R)\sqrt{\frac{C}{L}}.$$

Since at $x = 0$ we have $v(0, t) = f(t)$, then

$$V(0, s) = \mathcal{L}\{f(t)\} = F(s).$$

As we showed in (5.15), for an infinite line

$$V(x, s) = V(0, s)e^{-\gamma x}$$

and so

$$V(x, s) = F(s)e^{-\gamma x} = F(s)e^{-x(Ls+R)\sqrt{\frac{C}{L}}}.$$

From (5.3) we have, after Laplace transforming,

$$-\frac{dV(x, s)}{dx} = I(x, s)R + L[sI(x, s) - i(x, 0+)]$$

[9]The Atlantic Telegraph Cable operators didn't send traditional Morse code as dots and dashes, that is, as pulses of different *durations*, but rather used a special key to send equal duration pulses of opposite *polarity* — one polarity for a 'dot,' and the other polarity for a 'dash.'

and so, since $i(x, 0+) = 0$,

$$I(x,s) = -\frac{1}{Ls+R}\frac{dV(x,s)}{dx}$$

or,

$$I(x,s) = \sqrt{\frac{C}{L}}F(s)e^{-x(Ls+R)\sqrt{\frac{C}{L}}}$$

which means

$$I(x,s) = \sqrt{\frac{C}{L}}e^{-xR\sqrt{\frac{C}{L}}}F(s)e^{-x\sqrt{LC}s}. \tag{5.25}$$

Using (3.14) on (5.25), with

$$t_0 = x\sqrt{LC},$$

we can write

$$i(x,t) = \left\{\sqrt{\frac{C}{L}}e^{-xR\sqrt{\frac{C}{L}}}\right\}f\left(t - x\sqrt{LC}\right)u\left(t - x\sqrt{LC}\right). \tag{5.26}$$

Look carefully at (5.26). It says $i(x, t)$ is simply an attenuated (by the factor in the curly brackets), time-delayed (but otherwise perfect) *replica* of $f(t)$. The current at the point distance x from the sending end of the cable faithfully reproduces the *shape* of the input signal, with a time delay of $\sqrt{LC}$. We can think of the input signal as a wave traveling down the line at the speed $\frac{1}{\sqrt{LC}}$. (Look back at (5.9) and the related discussion there.)

To appreciate the numbers involved, suppose $R = 2$ ohms/nm, $C = 0.5$ microfarads/nm, and $G = 10^{-7}$ siemens/nm (that is, the cable leakage *resistance* is 10 megohms/nm). To satisfy Heaviside's distortionless condition means

$$L = \frac{RC}{G} = \frac{(2)\left(0.5 \times 10^{-6}\right)}{10^{-7}} = 1 \text{ henry/nm}.$$

This value of L is almost certainly larger than the inherent distributed inductance of a cable, and so L must be intentionally enhanced (by, for example, periodically inserting wire coils in the cable, a technique called *inductive loading*[10]). Putting

[10]In 1889 Thomson served as President of the Institution of Electrical Engineers and, in his Presidential Address, he heaped praise on Heaviside's discovery of inductive loading. Loading was of enormous practical value to the communications industry, and it made a lot of money for a number of people — but not for Heaviside. For an historical discussion of this sad tale, see the Heaviside biography (note 9 in the Preface).

these numbers into (5.26), we see the current at $x = 2,000$ nm, due to a step input of one volt at $x = 0$, is

$$\sqrt{\frac{C}{L}}e^{-xR\sqrt{\frac{C}{L}}} = \sqrt{0.5 \times 10^{-6}}e^{-2,000(3)\sqrt{0.5\times 10^{-6}}} \text{ amperes} = 101.6 \text{ microamperes}.$$

Thus, a 40 volt step input would produce a current at 2, 000 nm in excess of 4 milliamperes. The speed of propagation on the cable is

$$\frac{1}{\sqrt{LC}} = \frac{1}{\sqrt{0.5 \times 10^{-6}}} \text{ nm/second} = 1,414 \text{ nm/second}$$

and so current would start arriving at $x = 2,000$ nm after a delay of about 1.4 seconds.

5.5 The General, Infinite Transmission Line

Now, at last, we are ready to study the problem — first tackled by Heaviside but whose solution received posthumous skepticism (see note 6) — concerning the general (that is, arbitrary values for R, L, C, and G), initially relaxed infinite transmission line with a unit step voltage input. From (5.16), we start with

$$V(x, s) = \frac{1}{s}e^{-\gamma x}, = \sqrt{(Ls + R)(Cs + G)}. \tag{5.27}$$

What we mean by *solving* Heaviside's problem is using (5.27) to find the current $i(x, t)$ (the physical quantity that is detected at the receiving end of the line) for any given x. As we've done before, Laplace transforming (5.3), and using $i(x, 0+) = 0$, gives us

$$(R + Ls)I(x, s) = -\frac{dV(x, s)}{dx} + Li(x, 0+)$$

and so

$$I(x, s) = -\frac{1}{(R + Ls)}\frac{dV(x, s)}{dx} = -\frac{1}{(R + Ls)}\left[-\gamma\frac{1}{s}e^{-\gamma x}\right].$$

That is,

$$I(x, s) = \frac{\gamma}{s(R + Ls)}e^{-\gamma x} = \frac{\sqrt{(Ls + R)(Cs + G)}}{s(R + Ls)}e^{-\gamma x}$$

and so

$$I(x,s) = \frac{1}{s}\sqrt{\frac{Cs\ +\ G}{Ls+R}}e^{-\gamma x}. \tag{5.28}$$

We can get (5.28) into a form we can invert back to the time domain with just a bit more algebra on the radical, as follows.

$$\sqrt{\frac{Cs\ +\ G}{Ls+R}} = \sqrt{\frac{C\left(s+\frac{G}{C}\right)}{L\left(s+\frac{R}{L}\right)}} = \sqrt{\frac{C}{L}}\frac{s+\ \frac{G}{C}}{\sqrt{\left(s+\frac{R}{L}\right)\left(s+\frac{G}{C}\right)}} = \sqrt{\frac{C}{L}}\frac{s+\ \frac{G}{C}}{\sqrt{s^2+\left(\frac{R}{L}+\frac{G}{C}\right)s+\frac{RG}{LC}}}$$

$$= \sqrt{\frac{C}{L}}\frac{s+\ \frac{G}{C}}{\sqrt{(s+a)^2-b^2}}$$

where, of course,

$$s^2+2as+a^2-b^2 = s^2+\left(\frac{R}{L}+\frac{G}{C}\right)s+\frac{RG}{LC}.$$

It therefore follows that

$$2a = \frac{R}{L}+\frac{G}{C}$$

and

$$a^2-b^2 = \frac{RG}{LC},$$

and I'll let you solve for a and b to confirm that

$$a = \frac{1}{2}\left(\frac{R}{L}+\frac{G}{C}\right) \tag{5.29}$$

and

$$b = \frac{1}{2}\left(\frac{R}{L}-\frac{G}{C}\right). \tag{5.30}$$

Thus,

$$\sqrt{\frac{Cs\ +\ G}{Ls+R}} = \sqrt{\frac{C}{L}}\frac{s+\ \frac{G}{C}}{\sqrt{(s+a)^2-b^2}}$$

and so, looking back at (5.28), we have

$$I(x,s)=\sqrt{\frac{C}{L}}\left(\frac{1}{s}\right)\frac{s+\frac{G}{C}}{\sqrt{(s+a)^2-b^2}}e^{-\gamma x}. \tag{5.31}$$

As it stands, we are unable to invert (5.31) because γ, itself, is a function of s and we haven't developed a transform pair that can handle an $I(x, s)$ of the complexity of (5.31). If, on the other hand, we are willing to settle for the *input* current at $x = 0$, then (5.31) reduces to

$$I(0,s)=\sqrt{\frac{C}{L}}\left[\frac{1}{\sqrt{(s+a)^2-b^2}}+\frac{G}{Cs\sqrt{(s+a)^2-b^2}}\right],$$

and this we *can* invert. Here's how, in a two-step process.

First, *if* we knew the time function that pairs with

$$\frac{1}{\sqrt{s^2-b^2}}$$

then the time function that pairs with

$$\frac{1}{\sqrt{(s+a)^2-b^2}}$$

would be that same time function multiplied by e^{-at}, a general result we derived in (3.13). And we *do* know that time function: it's the modified Bessel function of the first kind of order zero because, from (3.48), we have

$$I_0(bt)\leftrightarrow\frac{1}{\sqrt{s^2-b^2}}$$

and so

$$e^{-at}I_0(bt)\leftrightarrow\frac{1}{\sqrt{(s+a)^2-b^2}}.$$

Second, from (3.7) we have the general pair

$$\int_0^t f(y)dy\leftrightarrow\frac{F(s)}{s},$$

which tells us that

$$\int_0^t e^{-ay} I_0(by)dy \leftrightarrow \frac{1}{s\sqrt{(s+a)^2 - b^2}}.$$

Applying these results to (5.31) gives us

$$i(0,t) = \sqrt{\frac{C}{L}}\left[e^{-at} I_0(bt) + \frac{G}{C}\int_0^t e^{-ay} I_0(by)dy\right] \tag{5.32}$$

where a and b are as given in (5.29) and (5.30), respectively.

(5.32) was easy to get, but your reaction to its calculation is probably not one of enthusiasm. After all, the whole point to a transmission line is what comes *out* at the *receiving* end, not what goes *in* at the *sending* end! So, let's go back to (5.31) and do what all engineers do when, for example, they run into an integral they don't know how to evaluate: they get hold of a really good table of integrals and look it up. And that's just what we'll do here, too (except, of course, we'll instead get hold of a really good table of Laplace transform pairs). If you do that you'll find the pair

$$I_0\left(p\sqrt{t^2 - k^2}\right)u(t-k) \leftrightarrow \frac{e^{-k\sqrt{s^2 - p^2}}}{\sqrt{s^2 - p^2}}, \tag{5.33}$$

where p and k are constants and $u(t - k)$ is a shifted step function that is (of course) zero for $t < k$. We'll use (5.33) as follows.

We have

$$\gamma = \sqrt{(Ls+R)(Cs+G)} = \sqrt{LC}\sqrt{\left(s+\frac{R}{L}\right)\left(s+\frac{G}{C}\right)}$$

$$= \sqrt{LC}\sqrt{s^2 + s\left(\frac{R}{L}+\frac{G}{C}\right) + \frac{RG}{LC}}$$

$$= \sqrt{LC}\sqrt{(s+a)^2 - b^2}$$

and so

$$I(x,s) = \sqrt{\frac{C}{L}}\left(\frac{1}{s}\right)\frac{s+\frac{G}{C}}{\sqrt{(s+a)^2 - b^2}}e^{-x\sqrt{LC}\sqrt{(s+a)^2 - b^2}}$$

or,

$$I(x,s) = \sqrt{\frac{C}{L}}\left[\frac{e^{-x\sqrt{LC}\sqrt{(s+a)^2 - b^2}}}{\sqrt{(s+a)^2 - b^2}} + \frac{G}{C}\left(\frac{1}{s}\right)\frac{e^{-x\sqrt{LC}\sqrt{(s+a)^2 - b^2}}}{\sqrt{(s+a)^2 - b^2}}\right]. \tag{5.34}$$

In the notation of (5.33),

$$k = x\sqrt{LC} \text{ and } p = b.$$

From (5.33) we have the pair

$$I_0\left(b\sqrt{t^2 - x^2LC}\right)u\left(t - x\sqrt{LC}\right) \leftrightarrow \frac{e^{-x\sqrt{LC}\sqrt{s^2-b^2}}}{\sqrt{s^2 - b^2}}$$

and so it immediately follows that

$$I_0\left(b\sqrt{t^2 - x^2LC}\right)e^{-at}u\left(t - x\sqrt{LC}\right) \leftrightarrow \frac{e^{-x\sqrt{LC}\sqrt{(s+a)^2-b^2}}}{\sqrt{(s+a)^2 - b^2}}.$$

In addition, the second term in the brackets of (5.34) pairs as

$$\frac{G}{C}\int_0^t I_0\left(b\sqrt{y^2 - x^2LC}\right)e^{-ay}u\left(y - x\sqrt{LC}\right)dy \leftrightarrow \frac{G}{C}\left(\frac{1}{s}\right)\frac{e^{-x\sqrt{LC}\sqrt{(s+a)^2-b^2}}}{\sqrt{(s+a)^2 - b^2}}.$$

Thus, *finally* (!), we have our general result for an initially relaxed, infinite line with a step input of one volt:

$$i(x,t) = \sqrt{\frac{C}{L}}\left[\begin{array}{c} I_0\left(b\sqrt{t^2 - x^2LC}\right)e^{-at}u\left(t - x\sqrt{LC}\right) \\ +\frac{G}{C}\int_0^t I_0\left(b\sqrt{y^2 - x^2LC}\right)e^{-ay}u\left(y - x\sqrt{LC}\right)dy \end{array}\right] \tag{5.35}$$

where a and b are given by (5.29) and (5.30), respectively. (Notice that (5.35) reduces to (5.32) if $x = 0$.)

The first thing that is obvious, *by inspection*, is that for a given x

$$i(x,t) = 0, t < x\sqrt{LC},$$

which simply states the physical fact that there can be no current at distance x until the current has had enough time to get there (at speed $\frac{1}{\sqrt{LC}}$). Once we've said *that*, however, the next step of producing a plot of $i(x, t)$ isn't trivial. Indeed, when I was a student in the 1950s/60s, producing such a plot would have been a bit of a challenge, as the numerical work involved would have been fairly tedious. With *MATLAB*, however, it's just a matter of a few minutes to write a computer code that does *all* of the grubby work. The result is Fig. 5.4, which shows the first twenty seconds of $i(3{,}500, t)$ for a cable with the parameters $R = 1.8$ ohms/nm, $L = 0.1$ henrys/nm, $C = 0.4$ microfarads/nm, and $\frac{1}{G} = 10^7$ ohms/nm. These numbers were picked for no

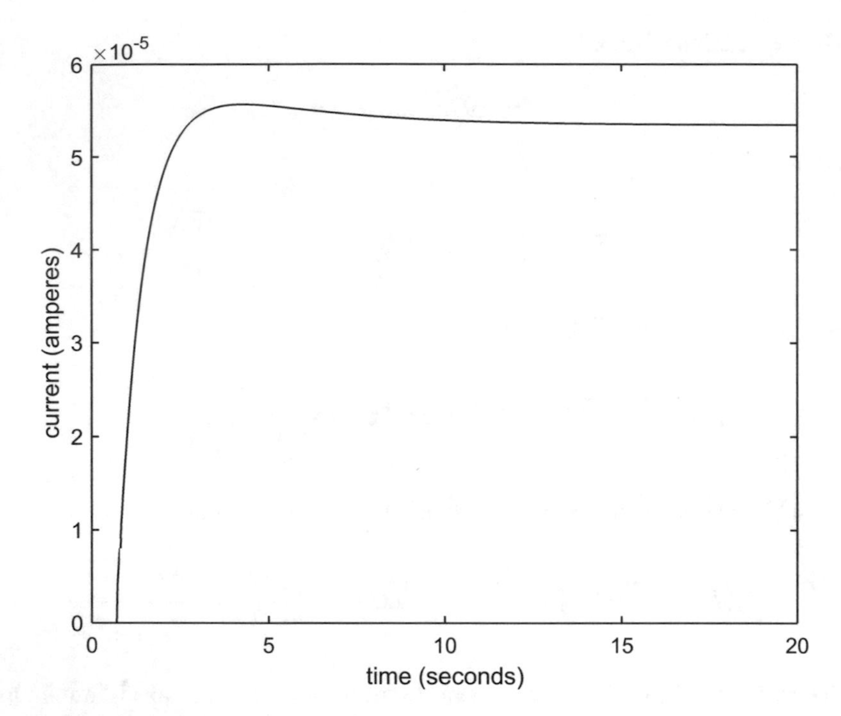

Fig. 5.4 The current at $x = 3, 500$ nm in a general cable

particular reason other than they were used in a paper[11] that suggested Heaviside's solution was incorrect. The result of (5.35) matches McLachlan's, but he did not use the Laplace transform (note the date of his paper). Rather, as you can tell from his title, he essentially performed a direct inversion in the complex plane of Heaviside's operational solution, a calculation that would have left most non-academic electrical engineers of his day in the dark.

The figure shows that, after a delay of 0.7 seconds (corresponding to traveling 3, 500 nm at a speed of 5, 000 nm/second), the current rises very quickly to about 55 microamperes (remember, for a one volt step input) and then (within two or three seconds) settles down to a steady value of just slightly less. We can, theoretically, calculate the precise value of that steady current from (5.35) by letting $t \to \infty$. The first term in the square brackets clearly goes to zero, but just what the second, integral term does isn't quite so clear. Much easier to do is to apply the final value theorem to $I(x, s)$ in (5.34). That is,

[11]N. W. McLachlan, "Submarine Cable Problems Solved By Contour Integration," *The Mathematical Gazette*, February 1938, pp. 37–41. There is no plot of $i(x, t)$ in this paper.

$$\lim_{t\to\infty} i(x,t) = \lim_{s\to 0} sI(x,s) = \sqrt{\frac{C}{L}}\left(\frac{G}{C}\right)\frac{e^{-x\sqrt{LC}\sqrt{a^2-b^2}}}{\sqrt{a^2-b^2}}.$$

I'll let you confirm, using the expressions for a and b given in (5.29) and (5.30), respectively, that

$$\lim_{t\to\infty} i(x,t) = \sqrt{\frac{G}{R}}e^{-x\sqrt{RG}} \tag{5.36}$$

and that, if you use $x = 3,500$, $R = 1.8$, and $G = 10^{-7}$, the result is 53.4 microamperes. This is in excellent agreement with Fig. 5.4.

To show you that we could have easily used any other set of numbers for the line's distributed parameters, the *MATLAB* code that created Fig. 5.4 (**general.m**) is as follows, about which I'll say nothing more because this is *not* a *MATLAB* coding text. The code executes in mere seconds, and is given here mostly to satisfy your possible curiosity, but I do want to direct your attention to the two appearances in it of the software step function *heaviside*, which The MathWorks (creator of *MATLAB*) included in *MATLAB* to honor Oliver Heaviside. He would almost surely have preferred money from the commercial cable telegraph companies for his discovery of the distortionless transmission line and inductive loading, but *MATLAB*'s post-humous recognition will have to do.

```
%general.m/created by PJNahin for Electrical Transients
(9/15/2017)
R=1.8;L=0.1;C=4e-7;G=1e-7;x=3500;
r1=R/L;r2=G/C;a=(r1+r2)/2;b=(r1-r2)/2;
delay=x*x*L*C;f=sqrt(C/L);speed=sqrt(delay);
t=linspace(0,20,1000);
for j=1:999
  t1(j)=(t(j)+t(j+1))/2;
end
arg=b*sqrt(t1.^2-delay);
term1=besseli(0,arg).*exp(-a*t1).*heaviside(t1-speed);
for j=1:999
  y=linspace(t(j),t(j+1),20);
  fun=besseli(0,b*sqrt(y.^2-delay)).*exp(-a*y).*heaviside
(y-speed);
  I(j)=trapz(y,fun);
end
I=cumsum(I);
current=f*(term1+r2*I);
plot(t1,current,'-k')
xlabel('time (seconds)')
ylabel('current (amperes)')
```

5.6 Transmission Lines of Finite Length

Not all lines are infinitely long. In fact, our assumption of an infinitely long transmission line is a convenient fiction that makes the math 'easier' to handle: in reality, there are *no* lines of truly infinite length.[12] So, what happens when we introduce the new restriction of finite length? The answer is that (5.15) is simply not true. So, let's back-up one step to (5.14) and rewrite it as

$$V(x,s) = K_1 e^{\gamma x} + K_2 e^{-\gamma x} \tag{5.37}$$

where

$$\gamma = \sqrt{(Ls + R)(Cs + G)}$$

and add recognition of the reality that $K_1 \neq 0$ (we argued earlier, you'll recall, that $K_1 = 0$ for an *infinite* line in order to keep $V(x, s)$ finite as $x \to \infty$). Let's now suppose that the line is actually of finite length d, and that the $x = 0$ end is *shorted*, as shown in Fig. 5.5, while at $x = d$ we switch a battery of voltage E in at $t = 0$. (If you're wondering why it is at $x = d$ that we apply the battery, rather than at $x = 0$, we *could* do that — after all, the result is simply to flip the line over in the reverse direction, which clearly changes nothing electrically — but the math would be a bit more awkward.)

Along with (5.37), we have two so-called *boundary conditions*, where 'boundary' refers to spatial constraints, rather than initial conditions which are temporal constraints:

$$v(0,t) = 0$$

and

$$v(d,t) = Eu(t).$$

These boundary conditions transform as

$$V(0,s) = 0$$

and

$$V(d,s) = \frac{E}{s}$$

[12]Does this mean that everything we've done so far with *infinite* lines has been a shaggy dog story (that is, just a big joke)? No. What our introduction of finite reality will do is give us a physical condition, *if satisfied*, that will allow us to treat a finite line as if it *were* actually infinite in length.

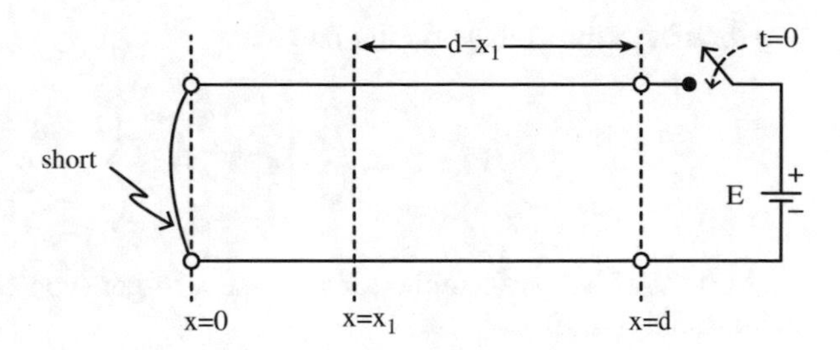

Fig. 5.5 A 'shorted' line of finite length

which, when applied to (5.37), give

$$K_1 + K_2 = 0$$

and

$$K_1 e^{\gamma d} + K_2 e^{-\gamma d} = \frac{E}{s}.$$

Clearly $K_1 = -K_2$ and substituting this into the last equation says

$$K_1 = \frac{E}{s(e^{\gamma d} - e^{-\gamma d})} = \frac{E}{2s\, sinh(\gamma d)}$$

and so

$$K_2 = -\frac{E}{2s\, sinh(\gamma d)}.$$

Substituting K_1 and K_2 back into (5.37), we thus have

$$V(x, s) = \frac{E}{2s\, sinh(\gamma d)}(e^{\gamma x} - e^{-\gamma x}) = \left(\frac{E}{s}\right)\frac{\sinh(\gamma x)}{\sinh(\gamma d)}. \tag{5.38}$$

Well, (5.38) *is* a solution, yes, but now you can see what I meant by saying earlier that the introduction of a finite length complicates the math, as now we have *hyperbolic* functions of γ! To go any further with (5.38) we'll need to start making some particular, simplifying assumptions, the easiest one of which is to assume that the line is lossless. That is, let's say $R = G = 0$, and so $\gamma = s\sqrt{LC}$. Then (5.38) becomes

$$V(x, s) = \left(\frac{E}{s}\right)\frac{e^{x\sqrt{LC}s} - e^{-x\sqrt{LC}s}}{e^{d\sqrt{LC}s} - e^{-d\sqrt{LC}s}} = \left(\frac{E}{s}\right)\frac{e^{x\sqrt{LC}s}}{e^{d\sqrt{LC}s}}\frac{\left(1 - e^{-2x\sqrt{LC}s}\right)}{\left(1 - e^{-2d\sqrt{LC}s}\right)}$$

or, if we write the propagation speed on the line as

$$c = \frac{1}{\sqrt{LC}},$$

we then have the voltage on the line as

$$V(x,s) = \left(\frac{E}{s}\right) e^{-s(d-x)/c} \frac{\left(1 - e^{-\frac{2x}{c}s}\right)}{\left(1 - e^{-\frac{2d}{c}s}\right)}. \tag{5.39}$$

Further, if we perform the long division to get (you can confirm this by simply cross-multiplying)

$$\frac{1}{\left(1 - e^{-\frac{2d}{c}s}\right)} = 1 + e^{-\frac{2d}{c}s} + e^{-\frac{4d}{c}s} + e^{-\frac{6d}{c}s} + \ldots$$

then

$$\begin{aligned} V(x,s) &= \left(\frac{E}{s}\right) e^{-s(d-x)/c} \left(1 - e^{-\frac{2x}{c}s}\right)\left[1 + e^{-\frac{2d}{c}s} + e^{-\frac{4d}{c}s} + e^{-\frac{6d}{c}s} + \ldots\right] \\ &= \left(\frac{E}{s}\right)\left(e^{-s(d-x)/c} - e^{-s(d+x)/c}\right)\left[1 + e^{-\frac{2d}{c}s} + e^{-\frac{4d}{c}s} + e^{-\frac{6d}{c}s} + \ldots\right] \end{aligned}$$

and so

$$V(x,s) = \left(\frac{E}{s}\right)\left[\begin{array}{l} \left\{e^{-\frac{s(d-x)}{c}} - e^{-\frac{s(d+x)}{c}}\right\} \\ +\left\{e^{-\frac{s(3d-x)}{c}} - e^{-\frac{s(3d+x)}{c}}\right\} \\ +\left\{e^{-\frac{s(5d-x)}{c}} - e^{-\frac{s(5d+x)}{c}}\right\} \\ +\left\{e^{-\frac{s(7d-x)}{c}} - e^{-\frac{s(7d+x)}{c}}\right\} \\ +\ldots \end{array}\right]. \tag{5.40}$$

At this point, (5.40) probably looks pretty terrifying, but it actually has a beautifully simple physical interpretation. When we return to the time domain, the $\left(\frac{E}{s}\right)$ factor tells us we are going to get step functions, step functions that are ever-more delayed in time because of the exponentials in the curly brackets (look back at (3.14)). That is,

$$v(x,t) = E\left[\begin{array}{l} \left\{u\left(t - \frac{d-x}{c}\right\} - \left\{u\left(t - \frac{d+x}{c}\right\}\right.\right. \\ +\left\{u\left(t - \frac{3d-x}{c}\right\} - \left\{u\left(t - \frac{3d+x}{c}\right\}\right.\right. \\ +\left\{u\left(t - \frac{5d-x}{c}\right\} - \left\{u\left(t - \frac{5d+x}{c}\right\}\right.\right. \\ +\left\{u\left(t - \frac{7d-x}{c}\right\} - \left\{u\left(t - \frac{7d+x}{c}\right\}\right.\right. \\ +\ldots \end{array}\right]$$

and Fig. 5.6 shows what $v(x, t)$ looks like at some arbitrary $0 < x = x_1 < d$. That figure looks pretty 'busy' for a simple step input, and the reason behind that is the short-circuit at $x = d$. Here's why.

When E is applied at $t = 0$, a wave with that amplitude starts propagating down the line at speed c, reaching $x = x_1$ at time $t = \frac{d-x_1}{c}$, and so we see the line voltage jump from zero to E at that instant. The wave continues on down the line, reaching the short circuit at time $= \frac{d}{c}$. The voltage at a *short* is, by definition, zero, and so we seem to have a problem, as the propagating wave has amplitude E. The way out of this puzzle is to imagine that, just as the wave reaches the short (let's call this the *incident* wave), a *reflected* wave of amplitude –E is encountered, cancelling the incident wave and in that way giving the required sum of zero at $x = d$. This echo wave travels back up the line towards the battery, reaching $x = x_1$ at time $t = \frac{d+x_1}{c}$, thus causing the line voltage at $x = x_1$ to drop back down to zero, just as shown in Fig. 5.6.

The echo wave with amplitude –E reaches $x = 0$ where it 'sees' a voltage of E (the battery), and so we again have an apparent problem. How do we make the echo wave of –E compatible with the encountered battery voltage of +E? Why, just as before: we now imagine a *second* echo wave, of amplitude E, propagating back down the line, which gives us the proper voltage of E at $x = d$. This echo wave reaches $x = x_1$ at $t = \frac{3d-x_1}{c}$ which sends the voltage at $x = x_1$ back up to E. And so it goes, forever, with reflected echo waves endlessly bouncing off both ends of the line (each bounce with a sign reversal), adding and subtracting at the times shown in Fig. 5.6. As long as no reflections have occurred on a transmission line of finite length, we can consider the line to be of *infinite* length. Once reflections occur, however, the finite length must be taken into account.

The voltage behavior as a function of time is indeed quite interesting, but I think you'll find what the current does even *more* non-intuitive. To find the current $i(x, t)$ for our finite length line, we do just as we've done before, using (5.3) with $R = 0$ to write (after transforming) and using (5.38),

$$I(x,s) = -\frac{1}{Ls}\frac{dV(x,s)}{dx} = -\frac{E\gamma}{Ls^2}\frac{\cosh(\gamma x)}{\sinh(\gamma d)},$$

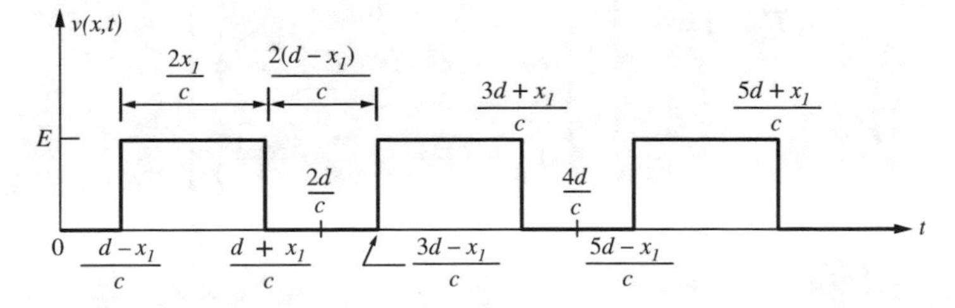

Fig. 5.6 How the voltage varies on a shorted, finite-length, lossless transmission line at distance x_1 from the shorted end

or, as $\gamma = \frac{s}{c}$, we have

$$I(x,s) = -\frac{E\frac{s}{c}}{Ls^2}\frac{\cosh\left(\frac{sx}{c}\right)}{\sinh\left(\frac{sd}{c}\right)} = -\frac{E}{Lcs}\frac{\cosh\left(\frac{sx}{c}\right)}{\sinh\left(\frac{sd}{c}\right)}$$

or, since $c = \frac{1}{\sqrt{LC}}$,

$$I(x,s) = -\frac{E}{s\sqrt{\frac{L}{C}}}\left(\frac{e^{\frac{sx}{c}} + e^{-\frac{sx}{c}}}{e^{\frac{sd}{c}} - e^{-\frac{sd}{c}}}\right).$$

In Appendix 2, the quantity $\sqrt{\frac{L}{C}}$ is called the *characteristic impedance* of the line (we see it's purely real, and so is actually a characteristic *resistance*), which we'll call R_0. Thus,

$$\begin{aligned} I(x,s) &= -\frac{E}{sR_0}\left(\frac{e^{\frac{sx}{c}} + e^{-\frac{sx}{c}}}{e^{\frac{sd}{c}} - e^{-\frac{sd}{c}}}\right) = -\frac{E}{sR_0}\left[\frac{e^{\frac{sx}{c}} + e^{-\frac{sx}{c}}}{e^{\frac{sd}{c}}\left(1 - e^{-2\frac{sd}{c}}\right)}\right] \\ &= -\frac{E}{sR_0}\left[\frac{e^{-s\frac{d-x}{c}} + e^{-s\frac{d+x}{c}}}{1 - e^{-2\frac{sd}{c}}}\right] \end{aligned}$$

and so, recalling our earlier encounter with expanding $\left(1 - e^{-2\frac{sd}{c}}\right)^{-1}$, we have

$$I(x,s) = -\frac{E}{sR_0}\left(e^{-s\frac{d-x}{c}} + e^{-s\frac{d+x}{c}}\right)\left[1 + e^{-2\frac{sd}{c}} + e^{-4\frac{sd}{c}} + e^{-6\frac{sd}{c}} + \ldots\right]$$

or,

$$I(x,s) = -\frac{E}{sR_0}\begin{bmatrix} \left\{e^{-s\frac{d-x}{c}} + e^{-s\frac{d+x}{c}}\right\} \\ + \\ \left\{e^{-s\frac{3d-x}{c}} + e^{-s\frac{3d+x}{c}}\right\} \\ + \\ \left\{e^{-s\frac{5d-x}{c}} + e^{-s\frac{5d+x}{c}}\right\} \\ + \\ \left\{e^{-s\frac{7d-x}{c}} + e^{-s\frac{7d+x}{c}}\right\} \\ + \ldots \end{bmatrix}. \tag{5.41}$$

$I(x, s)$ inverts back to the time domain to give

$$i(x,t) = -\frac{E}{R_0}\begin{bmatrix} \left\{u\left(t-\frac{d-x}{c}\right)+u\left(t-\frac{d+x}{c}\right)\right\} \\ + \\ \left\{u\left(t-\frac{3d-x}{c}\right)+u\left(t-\frac{3d+x}{c}\right)\right\} \\ + \\ \left\{u\left(t-\frac{5d-x}{c}\right)+u\left(t-\frac{5d+x}{c}\right)\right\} \\ + \\ \left\{u\left(t-\frac{7d-x}{c}\right)+u\left(t-\frac{7d+x}{c}\right)\right\} \\ +\ldots \end{bmatrix}. \tag{5.42}$$

As with the voltage, we get reflections for the current at each end of the line but, unlike the voltage in which these reflections *cancel*, now the reflections *add*. The result is shown in Fig. 5.7, which indicates that the current at $x = x_1$ (the minus sign in (5.42) appears because the direction of the line current in this analysis is opposite the positive current direction in Fig. 5.1) grows without bound (no surprise there as, after all, the line is shorted) *with discrete amplitude jumps at discrete instants of time* (this feature is probably *not* something you anticipated from your experience with how lumped parameter circuits behave when shorted).

Well, I think you can now appreciate how many new questions the introduction of a finite-length to a transmission line has prompted. What if the line is *open*-circuited? What if it is terminated in a finite, *non*-zero impedance? What if the line is not just a simple lossless line? What if . . .? With those tantalizing questions for you to ponder, this book has reached its end (or else it wouldn't be that *little* book for EE116 I told you about at the start). I think, however, that you've now got all the tools you need to move on into more advanced writings on these (and other) questions about the transient, switched behavior of finite lines.

So, off you go!

Problems

5.1 Take-up the challenge of note 7.

5.2 Establish (5.23) and (5.24).

5.3 Is the final value theorem consistent with (5.18) and (5.22)?

5.4 At one point in his 1855 paper (note 3), Thomson wrote of "putting a very intense battery [one of very high voltage] in communication with the end [of a submarine cable] from which the signal is sent, for a very short time, and then instantly putting this end in communication with the ground." That is, Thomson was talking of an impulsive voltage input to a transmission line (this was *seventy*

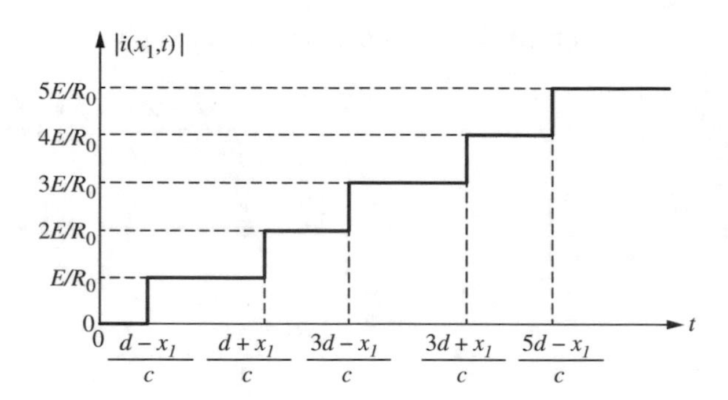

Fig. 5.7 How the current varies on a shorted, finite-length line at distance x_1 from the shorted end

years (!) before Dirac).[13] Calculate and plot $i(x, t)$ for the Atlantic Cable for $v(0, t) = \delta(t)$.

5.5 Establish (5.36).

5.6 Look back at (5.16), which assumes a step input to an infinite transmission line. If, instead, the input is an arbitrary time function $f(t)$, show that $V(x, s) = F(s)e^{-\gamma x}$. Then, assuming the Atlantic Cable, find $v(x, t)$. Hint: use the convolution theorem. If, after some thought, you're stuck, take a look at the two shaded boxes at the end of Appendix 3 (but don't do that unless you're *really* stuck).

5.7 Suppose that in Fig. 5.5 the switch and battery are replaced with a short before the first echo arrives at $x = d$. That is, *both* ends of the line are now shorted. Sketch $v(x_1, t)$ and $i(x_1, t)$.

[13] It is generally not a good idea to apply an impulse to anything. It makes for interesting mathematical exercises, yes, but in the real world it tends to break stuff. An actual historical example of this comes from the Atlantic Cable itself: the Chief Electrician on the third attempt to lay the cable (the first two failed) insisted on using an induction coil producing 2,000 volts, which resulted in the destruction of the cable's insulation. That Chief Electrician (who was *not* Thomson) was, not surprisingly, soon after dismissed.

Appendix 1: Euler's Identity[1]

$$e^{it} = \cos(t) + i\sin(t), i = \sqrt{-1}$$

There are numerous ways of establishing this identity, one of immense value to electrical engineers in the analysis of transients, but here's one of the easiest to follow. Suppose a time function $f(t)$ is written in the form of a power series. That is,

$$f(t) = c_0 + c_1 t + c_2 t^2 + \ldots + c_n t^n + \ldots .$$

It's a freshman calculus exercise to show that all the coefficients follow from the general rule

$$c_n = \frac{1}{n!}\left(\frac{d^n f}{dt^n} \mid _{t=0}\right), n \geq 1,$$

that is, by taking successive derivatives of $f(t)$ and, after each differentiation, setting $t = 0$. (The $n = 0$ case means, literally, *don't* differentiate, just set $t = 0$.) In this way it is found, for example, that

$$\sin(t) = t - \frac{1}{3!}t^3 + \frac{1}{5!}t^5 - \ldots,$$
$$\cos(t) = 1 - \frac{1}{2!}t^2 + \frac{1}{4!}t^4 - \ldots,$$
$$e^t = 1 + t + \frac{1}{2!}t^2 + \frac{1}{3!}t^3 + \frac{1}{4!}t^4 + \frac{1}{5!}t^5 + \ldots .$$

[1]Named after the Swiss-born mathematician Leonhard Euler (1707–1783), who first published the identity in a 1748 book (although there is evidence he knew it as early as 1740).

P. J. Nahin, *Transients for Electrical Engineers*,
https://doi.org/10.1007/978-3-319-77598-2

Now, in the last series, set $t = ix$. Then

$$\begin{aligned} e^{ix} &= 1 + ix + \frac{1}{2!}(ix)^2 + \frac{1}{3!}(ix)^3 + \frac{1}{4!}(ix)^4 + \frac{1}{5!}(ix)^5 + \dots \\ &= 1 + ix - \frac{1}{2!}x^2 - i\frac{1}{3!}x^3 + \frac{1}{4!}x^4 + i\frac{1}{5!}x^5 + \dots \\ &= \left(1 - \frac{1}{2!}x^2 + \frac{1}{4!}x^4 - \dots\right) + i\left(x - \frac{1}{3!}x^3 + \frac{1}{5!}x^5 + \dots\right) \\ &= \cos(x) + isin(x). \end{aligned}$$

This is an identity in x, and so continues to hold if we replace every x with a t to give us our result:

$$e^{it} = \cos(t) + i\ sin(t).$$

This result immediately tells us that

$$e^{-it} = e^{i(-t)} = \cos(-t) + i\ sin(-t) = \cos(t) - i\ sin(t).$$

Thus,

$$e^{it} + e^{-it} = 2\cos(t)$$

and

$$e^{it} - e^{-it} = 2i\sin(t)$$

and so

$$\cos(t) = \frac{e^{it} + e^{-it}}{2},\ \sin(t) = \frac{e^{it} - e^{-it}}{2i}.$$

Euler's identity tells us that whenever we encounter an exponential raised to an imaginary power in a transient analysis, we should be on the look-out for oscillatory (sinusoidal) behavior. (But not always; look at Fig. 1.6, for example, and the related discussion.) Euler's identity is the source of many almost magical results. Here's just one. If $t = \frac{\pi}{2}$ then $e^{i\frac{\pi}{2}} = \cos\left(\frac{\pi}{2}\right) + i\ sin\left(\frac{\pi}{2}\right) = i$. Thus, $\left\{e^{i\frac{\pi}{2}}\right\}^i = i^i = e^{i^2\frac{\pi}{2}} = e^{-\frac{\pi}{2}} \approx 0.2078\dots$. That is, an *imaginary* number raised to an *imaginary* power can be a *real* number. Who would have guessed it!?

That's pretty marvelous, alright, but in fact the situation is even *more* marvelous. Since, for n *any* integer, positive, negative, or zero, we have

$$e^{i2\pi n} = \cos(2\pi n) + i\sin(2\pi n) = 1,$$

then

$$e^{i\frac{\pi}{2}}e^{i2\pi n} = (i)(1) = i = e^{i\left(\frac{\pi}{2}+2\pi n\right)} = e^{i\pi\left(\frac{1}{2}+2n\right)}.$$

Thus,

$$i^i = \left\{e^{i\pi\left(\frac{1}{2}+2n\right)}\right\}^i = e^{i^2\pi\left(\frac{1}{2}+2n\right)} = e^{-\pi\left(\frac{1}{2}+2n\right)}$$

and so we see that an imaginary number raised to an imaginary power is equal to an *infinity* of real numbers! I feel safe in saying that *nobody* (before Euler) would have guessed *that*.

Another important result in mathematics, involving a sum of trigonometric functions, can be easily derived with Euler's identity. Consider the sum (which occurs in theoretical studies of the convergence behavior of Fourier series, a topic briefly discussed in Appendix 3)

$$2\sum_{n=1}^{N} \cos(n\phi) = 2\{\cos(\phi) + \cos(2\phi) + \ldots + \cos(N\phi)\}.$$

This 'looks' like it might have a fairly complicated plot and, for the case of $N = 7$ for example,

Figure A1.1 shows that guess isn't too far off the mark.

Now, from Euler's identity,

$$2\sum_{n=1}^{N} \cos(n\phi) = 2\sum_{n=1}^{N}\left\{\frac{e^{in\phi} + e^{-in\phi}}{2}\right\} = \sum_{n=1}^{N} e^{in\phi} + \sum_{n=1}^{N} e^{-in\phi}.$$

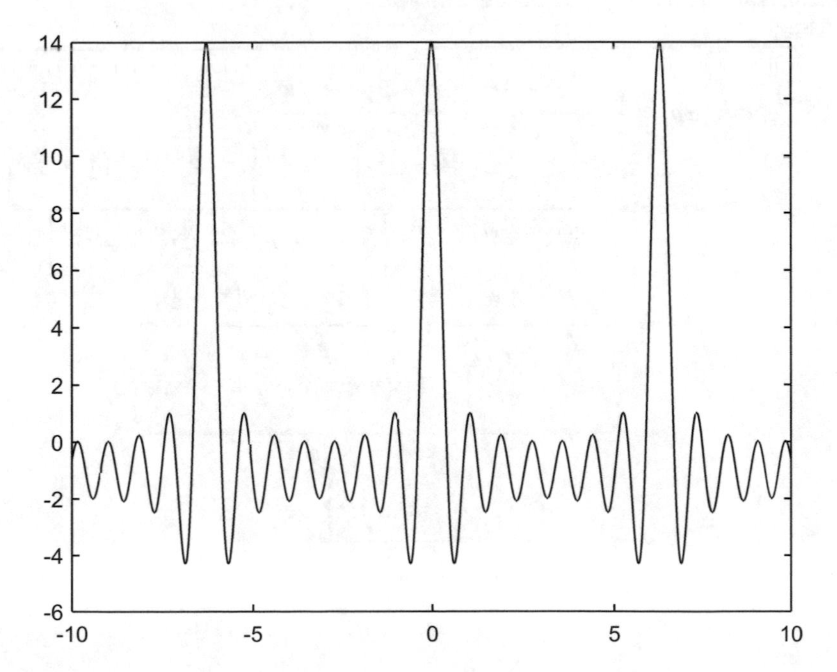

Fig. A1.1 The sum $2\sum_{n=1}^{7} \cos(n\phi)$

If we write

$$S = \sum_{n=1}^{N} e^{in\phi} = e^{i\phi} + e^{i2\phi} + \ldots + e^{iN\phi}$$

then

$$e^{i\phi}S = e^{i2\phi} + \ldots + e^{iN\phi} + e^{i(N+1)\phi}$$

and so

$$S - e^{i\phi}S = S\left(1 - e^{i\phi}\right) = e^{i\phi} - e^{i(N+1)\phi}$$

or,

$$S = \sum_{n=1}^{N} e^{in\phi} = \frac{e^{i\phi} - e^{i(N+1)\phi}}{1 - \ e^{i\phi}}.$$

In the same way,

$$\sum_{n=1}^{N} e^{-in\phi} = \frac{e^{-i\phi} - e^{-i(N+1)\phi}}{1 - \ e^{-i\phi}}.$$

You can see this last result, *instantly*, by simply recognizing this last sum is the previous sum with every ϕ replaced with $-\phi$.

Thus,

$$\begin{aligned}
2\sum_{n=1}^{N} \cos(n\phi) &= \frac{e^{i\phi} - e^{i(N+1)\phi}}{1 - \ e^{i\phi}} + \frac{e^{-i\phi} - e^{-i(N+1)\phi}}{1 - \ e^{-i\phi}} \\
&= \frac{e^{i\frac{1}{2}\phi}\left[e^{i\frac{1}{2}\phi} - e^{i\left(N + \frac{1}{2}\right)\phi}\right]}{e^{i\frac{1}{2}\phi}\left[e^{-i\frac{1}{2}\phi} - e^{i\frac{1}{2}\phi}\right]} + \frac{e^{-i\frac{1}{2}\phi}\left[e^{-i\frac{1}{2}\phi} - e^{-i\left(N + \frac{1}{2}\right)\phi}\right]}{e^{-i\frac{1}{2}\phi}\left[e^{i\frac{1}{2}\phi} - e^{-i\frac{1}{2}\phi}\right]} \\
&= \frac{e^{i\left(N + \frac{1}{2}\right)\phi} - e^{i\frac{1}{2}\phi}}{e^{i\frac{1}{2}\phi} - e^{-i\frac{1}{2}\phi}} + \frac{e^{-i\frac{1}{2}\phi} - e^{-i\left(N + \frac{1}{2}\right)\phi}}{e^{i\frac{1}{2}\phi} - e^{-i\frac{1}{2}\phi}} \\
&= \frac{\left[e^{-i\frac{1}{2}\phi} - e^{i\frac{1}{2}\phi}\right] + \left[e^{i\left(N + \frac{1}{2}\right)\phi} - e^{-i\left(N + \frac{1}{2}\right)\phi}\right]}{e^{i\frac{1}{2}\phi} - e^{-i\frac{1}{2}\phi}} \\
&= -1 + \frac{2i\sin\left\{\left(N + \frac{1}{2}\right)\phi\right\}}{2i\sin\left(\frac{1}{2}\phi\right)}
\end{aligned}$$

or, at last,

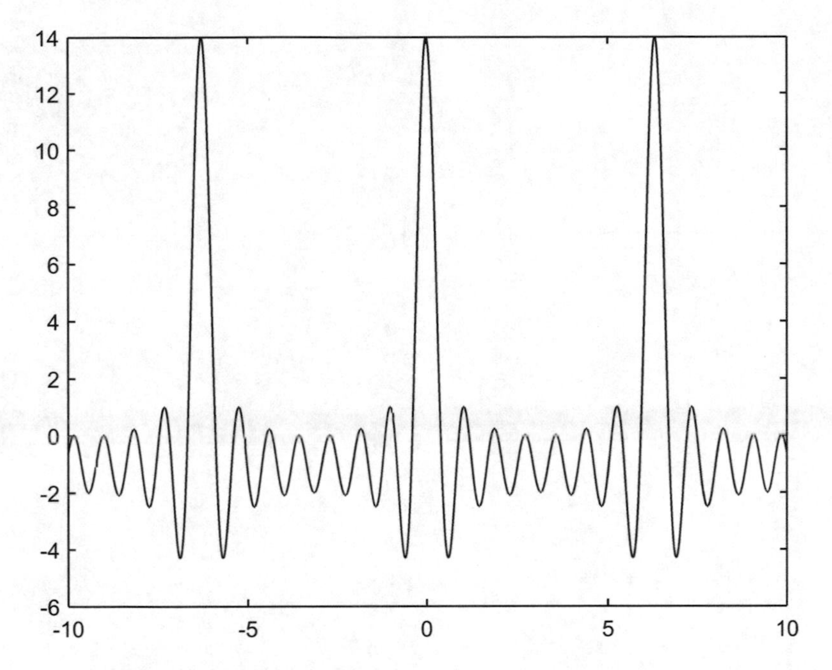

Fig. A1.2 A plot of $-1+\frac{sin\left\{\left(7\ +\ \frac{1}{2}\right)\phi\right\}}{sin\left(\frac{1}{2}\phi\right)}$

$$2\sum\nolimits_{n=1}^{N}\cos\left(n\phi\right)=-1+\frac{sin\left\{\left(N+\frac{1}{2}\right)\phi\right\}}{sin\left(\frac{1}{2}\phi\right)}.$$

The obvious question now is, what does the right-hand-side of *this* expression look like, and Fig. A1.2 shows that it looks *just like* (again, for $N = 7$) Fig. A1.1! The power of math (and Euler's identity). The two plots would, for *any* choice of *N*, be identical. Try it and see.

Appendix 2: Heaviside's Distortionless Transmission Line Condition

Heaviside gave, in 1887, a beautiful example of the utility of his resistance operator idea (look back at Problem 3.5) in his derivation of the condition for distortionless transmission on an infinitely long line. He imagined the line to be what is called a *periodic structure*. That is, as a circuit consisting of a series connection of an infinity of identical sub-circuits, as shown in Fig. A2.1. If we denote the conductance, capacitance, resistance, and inductance of the line *per unit length* by G, C, R, and L, respectively, then, if each sub-circuit represents a length l of the line, the interiors of all the sub-circuit boxes are as shown in the left-most box.[2]

Heaviside's great insight was that the impedance Z seen 'looking into' the line at the input of the first box is the same as the impedance that would be seen if the first box were discarded and one 'looked into' the input of the second box. As Heaviside wrote, "Since the circuit [the transmission line] is infinitely long, Z cannot be altered by cutting-off from the beginning ... any length."[3] In this way Heaviside reduced the infinite Fig. A2.1 to the finite Fig. A2.2.

Now, remember that Gl is a *conductance* and so represents a *resistance* of $\frac{1}{Gl}$, and the capacitance Cl represents an impedance of $\frac{1}{sCl}$. The right-most vertical series connection of Rl, Ll, and Z has an impedance of $Rl + sLl + Z$. The three vertical paths are in parallel, and the overall impedance of the parallel paths is Z. The *reciprocal* of

[2]The interior of the first box doesn't look quite like Figure 5.1.1, which was drawn as it is there for reasons of symmetry. Heaviside's arrangement has the same technical content, however, and it is a bit easier to use in deriving the condition for distortionless transmission.

[3]This idea is physically plausible and mathematically attractive — **but** — it has some quite subtle aspects to it that can lead the unwary astray. See, for example, S. J. van Enk, "Paradoxical Behavior of an Infinite Ladder Network," *American Journal of Physics*, September 2000, pp. 854–856. As for the unwary, even the great American mathematical physicist and Nobel prize winner Richard Feynman (1918–1988) may have not fully appreciated the subtleties: see my book *Mrs. Perkins's Electric Quilt*, Princeton 2009, pp. 24–36. These concerns do not cause trouble, fortunately, in Heaviside's analysis.

P. J. Nahin, *Transients for Electrical Engineers*,
https://doi.org/10.1007/978-3-319-77598-2

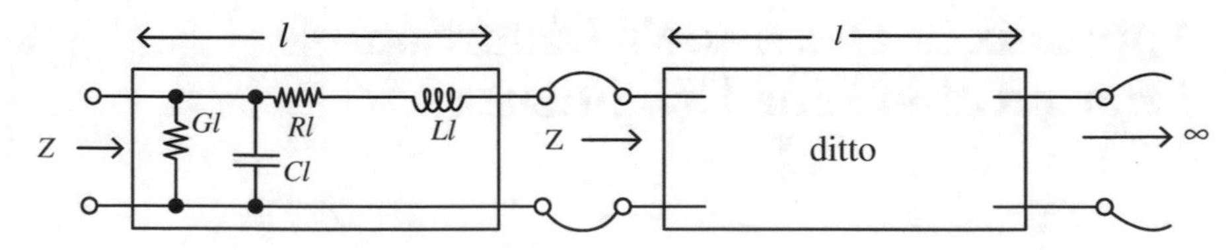

Fig. A2.1 An infinite transmission line as a series connection of an infinity of sub-circuits

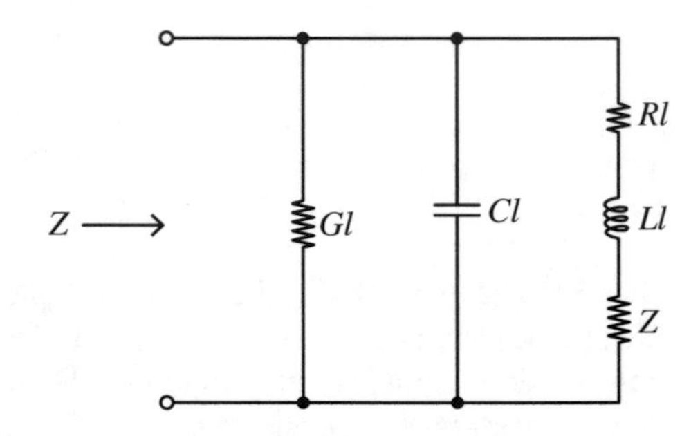

Fig. A2.2 This is Fig. A2.1 because the transmission line is *infinitely* long

the total impedance of the parallel paths is found by adding the *reciprocals* of the individual path impedances, and so Heaviside wrote

$$\frac{1}{Z} = \frac{1}{\frac{1}{Gl}} + \frac{1}{\frac{1}{sCl}} + \frac{1}{Rl + sLl + Z}$$

or

$$\frac{1}{Z} = Gl + sCl + \frac{1}{Rl + sLl + Z}.$$

Thus,

$$\frac{1}{Z} - \frac{1}{Rl + sLl + Z} = Gl + sCl$$

which, with just a couple of lines more of algebra becomes

$$Z^2 + Z(R + sL)l = \frac{R + sL}{G + sC}$$

Now, let $l \to 0$ which makes the sub-circuit approximation better and better. Thus,

$$Z = \sqrt{\frac{R + sL}{G + sC}} = \sqrt{\frac{L\left(s + \frac{R}{L}\right)}{C\left(s + \frac{G}{C}\right)}} = \sqrt{\frac{L}{C}\sqrt{\frac{s + \frac{R}{L}}{s + \frac{G}{C}}}}$$

Heaviside then observed that *if*

$$\frac{R}{L} = \frac{G}{C}$$

then we see the s-dependency vanish because the common factors in the numerator and denominator cancel. Z becomes simply a real number, the *resistance* $\sqrt{\frac{L}{C}}$ which is called the *characteristic impedance* of the transmission line. In this case the transmission line treats every frequency component present in the spectrum of a signal exactly like every other frequency component in the signal, and that's why the condition in the box is called the distortionless condition. In Chap. 5 a specific solution — see (5.26) — of the transmission line differential equations, under the assumption of $RC = GL$, elaborates on this claim.

Appendix 3: How to Solve for the Step Response of the Atlantic Cable Diffusion Equation *Without* the Laplace Transform

In Chap. 5 we studied Thomson's diffusion equation (5.10) for the leakage-free, non-inductive cable,

$$\frac{\partial^2 v}{\partial x^2} = RC\frac{\partial v}{\partial t},$$

or, to put it in the form you'll find in math text books,

$$\frac{\partial v}{\partial t} = k\frac{\partial^2 v}{\partial x^2}, k = \frac{1}{RC} \tag{A3.1}$$

which we solved using the Laplace transform. If you look at the writings of both Fourier and Thomson, however, you'll not find anything at all like Chap. 5. Instead, they used 'traditional,' classical mathematics. In this appendix I'll show you how that can be done (in addition to being a good history lesson on how the pioneers in transient analysis didn't let the lack of transform theory get in their way, what we'll do here will *really* show you why you should appreciate the transform!).

The mathematics involved is not normally encountered by an electrical engineering student until a first course in partial differential equations, but that's not actually an indication of advanced difficulty. Yes, some of the arguments I'll use here may be novel to a student who has just finished the freshman year of college, but they are nonetheless *completely* accessible to such a student. Those arguments are not difficult to absorb, they are just *new*.

We start with an idea that has been traced back to as long ago as 1753, to the Swiss mathematician Daniel Bernoulli (1700–1782). We'll *assume* that the solution to (A3.1) can be written in the form of

$$v(x,t) = X(x)T(t). \tag{A3.2}$$

That is, we'll *assume* that we can separate the variables x and t into the product of two functions, $X(x)$ and $T(t)$. $X(x)$ has no t-dependence, and $T(t)$ has no

P. J. Nahin, *Transients for Electrical Engineers*,
https://doi.org/10.1007/978-3-319-77598-2

x-dependence. *Why* can we assume this? Well, we can assume anything we want in mathematics — the proof-in-the-pudding is if the assumption leads to a solution that works! This product assumption will pass that pragmatic test. Well, you ask, how did Bernoulli know to do this? Well, I answer, I'm not absolutely sure, but see the following shaded box for a speculation.

Daniel Bernoulli was a member of a celebrated *family* of mathematicians. In particular, his father was Jean Bernoulli (1669–1748), often known instead by the Anglicized form of John (or the German equivalent, Johann). Daniel's uncle (Jean's brother) was Jacques Bernoulli (1654–1705) who was often called James or Jacob. Here I'll use John and Jacob. Both men were highly talented mathematicians, but they were also a jealous, combative pair. In 1697 Jacob had been trying for months, unsuccessfully, to solve the differential equation $\frac{dy}{dx} = y(x)f(x) + y^n(x)g(x)$ where $f(x)$ and $g(x)$ are two given, arbitrary functions. When John learned of his brother's failure, his spirits soared (no brotherly love here!): after all, what better way to stick a thumb in his rival's eye than to solve what Jacob couldn't, which John promptly did by making the substitution $y(x) = u(x)v(x)$. That is, John *assumed* $y(x)$ is the product of two (as yet unknown) functions, $u(x)$ and $v(x)$. Substituting this into the differential equation gives $v\left(\frac{du}{dx}\right) + u\left(\frac{dv}{dx}\right) = v(uf) + u(u^{n-1}v^n g)$ and so, making the obvious associations across the equals sign, $\frac{du}{dx} = uf$ and $\frac{dv}{dx} = u^{n-1}v^n g$. Both of these equations are separable and so can be directly integrated. That is, $\frac{du}{u} = fdx$ and $\frac{dv}{v^n} = u^{n-1}gdx$ and thus (I'll let you finish the details) $u(x) = e^{\int_0^x f(t)dt}$, $v(x) = \left\{C + (1-n)\int_0^x u^{n-1}(t)g(t)f(t)dt\right\}^{\frac{1}{1-n}}$ where C is an arbitrary constant of integration. So clever was this idea that in August of 1697 the French mathematician Pierre Varignon (1654–1722) wrote to John to correctly say "In truth, there is nothing more ingenious than the solution that you give for your brother's equation, and this solution is so simple that one is surprised at how difficult the problem appeared to be: this is indeed what one calls an elegant solution." Daniel was certainly aware in 1753 of his father's nearly sixty year old trick, and so perhaps that was the inspiration for *his* product assumption. Well, okay, you say, but how did *John* know to make *his* assumption? Sorry, I'm all out of speculations!

Substituting (A3.2) into (A3.1), we get

$$\frac{1}{kT}\frac{dT}{dt} = \frac{1}{X}\frac{d^2X}{dx^2} \tag{A3.3}$$

The left-hand-side is a function of only t with no x-dependency, and the right-hand-side is a function of only x with no t-dependency, and this can be true for all

x and all t *only* if both sides are equal to the *same constant*. I'll write that constant as $-\lambda^2$, where λ is arbitrary.[4] Setting both sides of (A3.3) equal to $-\lambda^2$, we then have

$$\frac{dT}{dt} + k\lambda^2 T = 0 \tag{A3.4}$$

and

$$\frac{d^2X}{dx^2} + \lambda^2 X = 0. \tag{A3.5}$$

The general solutions to (A3.4) and (A3.5) are easily verified to be (by direct substitution)

$$X(x) = C_1 \cos(\lambda x) + C_2 \sin(\lambda x)$$

and

$$T(t) = C_3 e^{-\lambda^2 kt}$$

where C_1, C_2, and C_3 are arbitrary constants. Thus,

$$v(x,t) = C_3 e^{-\lambda^2 kt}[C_1 \cos(\lambda x) + C_2 \sin(\lambda x)]$$

or, combining constants in the obvious way to arrive at the constants A and B,

$$v(x,t) = Ae^{-\lambda^2 kt} \cos(\lambda x) + Be^{\lambda^2 kt} \sin(\lambda x). \tag{A3.6}$$

Now, before going any further, we need to specify the *boundary* and the *initial* conditions we are going to impose on $v(x, t)$. We are going to solve for $v(x, t)$ when a unit step voltage is applied at the $x = 0$ end of the cable. That is,

$$v(0,t) = 1, t > 0. \tag{A3.7}$$

Further, we'll suppose the cable is initially without any electrical charge, that is,

$$v(x,0) = 0, x > 0. \tag{A3.8}$$

Alas, if we try to use these two conditions on (A3.6), in an attempt to learn more about A, B, and λ, we immediately run into difficulties (give it a try).

To get around this difficulty, let me now show you a clever trick.[5] What we'll do is solve the diffusion equation using a *different* choice of boundary and initial conditions, and then I'll show you the easy connection this solution has to the

---

[4]Writing the constant as $-\lambda^2$ forces the constant to be *negative*. Now, here's a little math experiment for you: write the constant as λ^2 (which of course is always *positive*) and see what happens.

[5]Just to be sure there is no misunderstanding, this clever trick is *not* due to me! I should be so clever.

solution of the problem we are actually interested in. So, instead of (A3.7) and (A3.8), let's imagine a cable that has been connected, at $x = 0$, to a unit voltage for a very long time (*so long*, in fact, that the cable is fully charged), and then at $t = 0$ we *ground* the $x = 0$ end. Thus,

$$v(0,t) = 0, t > 0 \tag{A3.9}$$

and

$$v(x,0) = 1, x > 0 \tag{A3.10}$$

Keep in mind that (A3.6) is still valid, as it hasn't yet been constrained by any boundary or initial conditions.

When we apply (A3.9) to (A3.6) we get

$$Ae^{-\lambda^2 kt} = 0$$

which, for this to be true for all t, immediately tells us that $A = 0$. Thus,

$$v(x,t) = Be^{-\lambda^2 kt} \sin(\lambda x). \tag{A3.11}$$

Since λ is arbitrary, then (A3.11) holds for *all possible* choices for λ (since λ is *squared*, this means $0 < \lambda < \infty$ since using a negative value for λ adds nothing new). Thus, since the sum of two solutions to the diffusion equation is also a solution, then if we add terms like (A3.11) for *all possible* λ, that is, if we integrate over all non-negative λ, we will have the most general solution. Further, for each choice of λ, B itself could be a different constant. (The word *constant* simply means λ and B do not depend on either x or t.) That is, $B = B(\lambda)$. So, the *most* general solution is

$$v(x,t) = \int_0^\infty B(\lambda)e^{-\lambda^2 kt} \sin(\lambda x)d\lambda. \tag{A3.12}$$

We can find $B(\lambda)$ by applying (A3.10), which results in

$$v(x,0) = 1 = \int_0^\infty B(\lambda) \sin(\lambda x)d\lambda. \tag{A3.13}$$

The question now is, how do we solve (A3.13) for $B(\lambda)$, which is inside an integral? To answer this, using just the mathematics of Fourier's and Thomson's day, let's take a temporary break from the diffusion equation and indulge in a little digression into Fourier series. Imagine that $f(x)$ is some (any) periodic function with period T, that is, $f(x) = f(x + T)$. Then

$$f(x) = \frac{1}{2}a_0 + \sum_{n=1}^{\infty} \{a_n \cos(n\omega_0 x) + b_n \sin(n\omega_0 x)\}$$

where what is called the *fundamental frequency* is given by

$$\omega_0 = \frac{2\pi}{T}$$

The so-called *Fourier coefficients* are given by

$$a_n = \frac{2}{T}\int_{-T/2}^{T/2} f(x)\cos(n\omega_0 x)dx, \quad n = 0, 1, 2, 3, \ldots$$

and

$$b_n = \frac{2}{T}\int_{-T/2}^{T/2} f(x)\sin(n\omega_0 x)dx, \quad n = 1, 2, 3, \ldots$$

I'm not going to prove any of this (look in any good math book on Fourier series[6]), and will simply ask you to accept that the mathematicians have, indeed, established these statements.

Now, let's write $T = 2l$, and so the Fourier coefficients become

$$a_n = \frac{1}{l}\int_{-l}^{l} f(x)\cos\left(\frac{n\pi x}{l}\right)dx, \quad n = 0, 1, 2, 3, \ldots$$

and

$$b_n = \frac{1}{l}\int_{-l}^{l} f(x)\sin\left(\frac{n\pi x}{l}\right)dx, \quad n = 1, 2, 3, \ldots,$$

because $\omega_0 T = 2\pi$ says that, with $T = 2l$, we have

$$\omega_0 = \frac{\pi}{l}$$

The Fourier series for $f(x)$ is, then,

$$f(x) = \frac{1}{2}a_0 + \sum_{n=1}^{\infty}\left\{a_n\cos\left(\frac{n\pi x}{l}\right) + b_n\sin\left(\frac{n\pi x}{l}\right)\right\}.$$

Inserting our expressions for a_n and b_n, we have (with u as a dummy variable of integration),

[6]An excellent choice, in my opinion, is Georgi Tolstov, *Fourier Series*, Dover 1976. Tolstov (1911–1981) was a well-known mathematician at Moscow State University, and the author of numerous acclaimed math books. *Fourier Series* was one of his best.

$$f(x) = \frac{1}{2l}\int_{-l}^{l} f(u)du + \sum_{n=1}^{\infty} \cos\left(\frac{n\pi x}{l}\right)\frac{1}{l}\int_{-l}^{l} f(u)\cos\left(\frac{n\pi u}{l}\right)du$$
$$+ \sum_{n=1}^{\infty} \sin\left(\frac{n\pi x}{l}\right)\frac{1}{l}\int_{-l}^{l} f(u)\sin\left(\frac{n\pi u}{l}\right)du$$

or,

$$f(x) = \frac{1}{2l}\int_{-l}^{l} f(u)du$$
$$+ \sum_{n=1}^{\infty} \frac{1}{l}\int_{-l}^{l} f(u)\left\{\cos\left(\frac{n\pi x}{l}\right)\cos\left(\frac{n\pi u}{l}\right) + \sin\left(\frac{n\pi x}{l}\right)\sin\left(\frac{n\pi u}{l}\right)\right\}du.$$

If you now recall the identity

$$\cos(\alpha)\cos(\beta) + \sin(\alpha)\sin(\beta) = \cos(\alpha - \beta),$$

then with

$$\alpha = \frac{n\pi u}{l}, \beta = \frac{n\pi x}{l}$$

we have

$$f(x) = \frac{1}{2l}\int_{-l}^{l} f(u)du + \sum_{n=1}^{\infty} \frac{1}{l}\int_{-l}^{l} f(u)\,cos\left\{\frac{\pi n}{l}(u - x)\right\}du. \tag{A3.14}$$

Now, let $l \to \infty$, which means we have a periodic function whose period is the *entire* x-axis! In other words, $f(x)$ is now *any* function we wish.

What happens on the right-hand-side of (A3.14) as $l \to \infty$? The first thing we can say is, if $f(x)$ is an integrable function (the only kind that interest engineers studying real, physically realizable systems), then it bounds finite area and so

$$\lim_{l\to\infty} \frac{1}{2l}\int_{-l}^{l} f(u)du = 0.$$

Next, define

$$\lambda = \frac{\pi}{l}$$

and then write $\lambda_1 = \lambda$, $\lambda_2 = 2\lambda = \frac{2\pi}{l}$, $\lambda_3 = 3\lambda = \frac{3\pi}{l}$, ..., $\lambda_n = n\lambda = \frac{n\pi}{l}$, ... and so on, forever, as $n \to \infty$. If we write

$$\Delta\lambda_n = \lambda_{n+1} - \lambda_n = \frac{\pi}{l}$$

we have

$$\frac{1}{l} = \frac{\Delta\lambda_n}{\pi}$$

and so (A3.14) becomes (where I've dropped the first integral because we've agreed that it vanishes in the limit $l \to \infty$)

$$f(x) = \sum\nolimits_{n=1}^{\infty} \frac{\Delta\lambda_n}{\pi} \int_{-l}^{l} f(u)\, cos\, \{\lambda_n(u-x)\} du.$$

As $l \to \infty$ we see that $\Delta\lambda_n \to 0$, that is, $\Delta\lambda_n$ becomes ever smaller (ever more like the *differential* $d\lambda$), λ_n becomes the *continuous* variable λ, and the sum becomes an integral with respect to λ. (Mathematicians will cringe at this sort of talk, but engineers will at least give it a chance.) Since the definition of λ restricts it to non-negative values, we thus write the $l \to \infty$ limit as

$$\begin{aligned} f(x) &= \frac{1}{\pi} \int_0^{\infty} \left[\int_{-\infty}^{\infty} f(u)\, cos\, \{\lambda(u-x)\} du \right] d\lambda \\ &= \frac{1}{\pi} \int_0^{\infty} \left[\int_{-\infty}^{\infty} f(u) \{ \cos(\lambda u) \cos(\lambda x) + \sin(\lambda u) \sin(\lambda x) \} du \right] d\lambda \end{aligned}$$

where I've again used the identity $\cos(\alpha) \cos(\beta) + \sin(\alpha) \sin(\beta) = \cos(\alpha - \beta)$.

So, if we change notation and write $v(x, 0)$ instead of $f(x)$, just to make things look as we left (A3.13) when we started this digression, we can write

$$\begin{aligned} v(x,0) &= \frac{1}{\pi} \int_0^{\infty} \cos(\lambda x) \left\{ \int_{-\infty}^{\infty} v(u,0) \cos(\lambda u) du \right\} d\lambda \\ &+ \frac{1}{\pi} \int_0^{\infty} \sin(\lambda x) \left\{ \int_{-\infty}^{\infty} v(u,0) \sin(\lambda u) du \right\} d\lambda. \end{aligned} \tag{A3.15}$$

We know $v(x, 0) = 1$ for $x > 0$, while what $v(x, 0)$ is doing for $x < 0$ has *no physical significance* (the cable doesn't exist for $x < 0$). That means we can feel free to specify $v(x, 0)$ for $x < 0$ in any way we wish that's convenient. In particular, suppose we define $v(x, 0) = -1$ for $x < 0$, that is, $v(x, 0)$ is an *odd* function of x. Since $\cos(\lambda u)$ is an even function of u, and since $\sin(\lambda u)$ is an odd function of u, and since (by our recent definition) $v(u, 0)$ is an odd function of u, then

$$\int_{-\infty}^{\infty} v(u,0) \cos(\lambda u) du = 0$$

and

$$\int_{-\infty}^{\infty} v(u,0) \sin(\lambda u) du = 2 \int_0^{\infty} v(u,0) \sin(\lambda u) du.$$

Thus, (A3.15) becomes

$$1=\int_0^\infty \sin(\lambda x)\left\{\frac{2}{\pi}\int_0^\infty v(u,0)\sin(\lambda u)du\right\}d\lambda$$

or, as $v(u,0)=1$ in the inner integral as the dummy variable u varies from 0 to ∞, we have

$$1=\int_0^\infty \left\{\frac{2}{\pi}\int_0^\infty \sin(\lambda u)du\right\}\sin(\lambda x)d\lambda. \tag{A3.16}$$

Now, if you haven't noticed it yet, we have just found $B(\lambda)$! To see this, compare (A3.16) with (A3.13), and *now* you see it, don't you?:

$$B(\lambda)=\frac{2}{\pi}\int_0^\infty \sin(\lambda u)du.$$

Inserting the $B(\lambda)$ into (A3.12), we have[7]

$$v(x,t)=\int_0^\infty \left\{\frac{2}{\pi}\int_0^\infty \sin(\lambda u)du\right\}e^{-\lambda^2 kt}\sin(\lambda x)d\lambda$$

or, reversing the order of integration,

$$v(x,t)=\frac{2}{\pi}\int_0^\infty \left\{\int_0^\infty \sin(\lambda u)\sin(\lambda x)e^{-\lambda^2 kt}d\lambda\right\}du. \tag{A3.17}$$

If you recall the identity

$$\sin(\alpha)\sin(\beta)=\frac{1}{2}[\cos(\alpha-\beta)-\cos(\alpha+\beta)],$$

then (A3.17) becomes

$$\begin{aligned}v(x,t)=&\frac{1}{\pi}\int_0^\infty \left\{\int_0^\infty \cos\{\lambda(u-x)\}e^{-\lambda^2 kt}d\lambda\right\}du\\&-\frac{1}{\pi}\int_0^\infty \left\{\int_0^\infty \cos\{\lambda(u+x)\}e^{-\lambda^2 kt}d\lambda\right\}du.\end{aligned} \tag{A3.18}$$

[7]At this point, a perceptive reader might hesitate at the sight of the integral $\int_0^\infty \sin(\lambda u)du$. That's because for an engineer this integral is the area under the sine curve, over an infinite interval. What's that area? Is it zero? Couldn't it be *anything* from zero to $\frac{2}{\lambda}$? I think that the best way to think of this expression for $B(\lambda)$ is as a *symbolic* one, and to simply move right along to (A3.17).

The inner integrals of (A3.18) can be found by the simple expedient of using a good math table[8]:

$$\int_0^{\infty} e^{-ap^2} \cos(bp)dp = \frac{1}{2}\sqrt{\frac{\pi}{a}} e^{-b^2/4a}.$$

So, with $p = \lambda$, $a = kt$, and $b = u \pm x$, we have

$$\int_0^{\infty} \cos\{\lambda(u - x)\} e^{-\lambda^2 kt} d\lambda = \frac{1}{2}\sqrt{\frac{\pi}{kt}} e^{-(u-x)^2/4kt}$$

and

$$\int_0^{\infty} \cos\{\lambda(u + x)\} e^{-\lambda^2 kt} d\lambda = \frac{1}{2}\sqrt{\frac{\pi}{kt}} e^{-(u+x)^2/4kt}$$

which says (A3.18) becomes

$$v(x,t) = \frac{1}{2}\frac{1}{\sqrt{\pi kt}}\left[\int_0^{\infty} e^{-(u-x)^2/4kt} du - \int_0^{\infty} e^{-(u+x)^2/4kt} du\right]. \tag{A3.19}$$

Next, change variable in the two integrals of (A3.19) to

$$y = \frac{u \pm x}{2\sqrt{kt}}$$

where we use the minus sign in the first integral, and the plus sign in the second integral. Then,

$$v(x,t) = \frac{1}{2}\frac{1}{\sqrt{\pi kt}}\left[\int_{-\frac{x}{2\sqrt{kt}}}^{\infty} e^{-y^2} 2\sqrt{kt}\, dy - \int_{\frac{x}{2\sqrt{kt}}}^{\infty} e^{-y^2} 2\sqrt{kt}\, dy\right] = \frac{1}{\sqrt{\pi}} \int_{-\frac{x}{2\sqrt{kt}}}^{\frac{x}{2\sqrt{kt}}} e^{-y^2} dy$$

or, as e^{-y^2} is even about $y = 0$,

$$v(x,t) = \frac{2}{\sqrt{\pi}} \int_0^{\frac{x}{2\sqrt{kt}}} e^{-y^2} dy = erf\left(\frac{x}{2\sqrt{kt}}\right) \tag{A3.20}$$

as a look back at (3.78), the definition of the error function *erf*, will confirm.

[8]This definite integral has been known for centuries. It appears, for example, in Fourier's *Analytical Theory*, along with a derivation. If you are interested in the details of how such an integral can be attacked, see my book *Inside Interesting Integrals*, Springer 2015, pp. 77–79.

Now, remember that the $v(x, t)$ in (A3.20) is *not* the solution to the problem we are actually trying to solve. The solution in (A3.20) satisfies the conditions given in (A3.9) and (A3.10), while what we want is the solution that satisfies the conditions of (A3.7) and (A3.8). But that's *easy* to arrange — just subtract the $v(x, t)$ of (A3.20) from one! That is, our final answer is, with $k = 1/RC$,

$$v(x,t) = 1 - erf\left(\frac{x}{2\sqrt{kt}}\right) = 1 - erf\left(\frac{x}{2}\sqrt{\frac{RC}{t}}\right) \tag{A3.21}$$

which is, indeed, the solution we found in (5.17) using the Laplace transform. This works because 1 is a (trivial[9]) solution to the diffusion equation, and the difference of two solutions is also a solution.

Finally, to get the current in the Atlantic cable due to a unit voltage step input at $x = 0$, simply recall (5.3) with $L = 0$ (the Atlantic cable was, you'll remember, taken to be non-inductive):

$$-\frac{\partial v}{\partial x} = iR$$

Thus,

$$i(x,t) = -\frac{1}{R}\frac{\partial}{\partial x}\left[1 - \frac{2}{\sqrt{\pi}}\int_0^{\frac{x}{2\sqrt{kt}}} e^{-y^2}dy\right], k = 1/RC,$$

and the integral is easily differentiated with respect to x using Leibniz's formula (note 3 in Chap. 1). The result (which you should confirm) is

$$i(x,t) = \sqrt{\frac{C}{\pi Rt}}e^{-x^2RC/4t},$$

in agreement with (5.22) that was found with the aid of the Laplace transform.

Doing all of these calculations strictly in the time domain has been a *lot* of work. The transform approach, once you've gotten over the learning curve, is a *lot* easier, and I believe Fourier and Thomson would have loved it. As an illustration for why I say that, take a look back at Problem 5.6, which asks you to generalize the analysis for the response of the Atlantic cable to an *arbitrary* input at $x = 0$, and then read the following shaded box which uses the transform to make short work of the task.

[9] *Trivial*, because it reduces the diffusion equation to the claim that $0 = 0$ which, while undeniably true, is not of very much use!

Look back at (5.16), and how we derived there, using the Laplace transform, the voltage response of the Atlantic cable to a voltage step input. Instead of $(0, s) = \frac{1}{s}$, however, let's write $V(0, s) = F(s)$, where $F(s)$ is the transform of an *arbitrary* $f(t)$, applied to the $x = 0$ end of the cable. Then, $V(x, s) = F(s)$ $e^{-x\sqrt{RCs}}$ and so by a direct application of the convolution theorem — see (3.87) — we can immediately write

$$v(x, t) = f(t) * \mathcal{L}^{-1}\left\{e^{-x\sqrt{RCs}}\right\}$$

where $\mathcal{L}^{-1}$ denotes the inverse transform (that is, a function of time). From (3.85) we have

$$\frac{ae^{-\frac{a^2}{4t}}}{2\sqrt{\pi t^3}u(t) \leftrightarrow e^{-a\sqrt{s}}}$$

and so, with $a = x\sqrt{RC}$, we have $\mathcal{L}^{-1}\left\{e^{-x\sqrt{RCs}}\right\} = \frac{x\sqrt{RC}}{2\sqrt{\pi t^3}}e^{-x^2RC/4t}$. Thus,

$$v(x, t) = \frac{x}{2}\sqrt{\frac{RC}{\pi}}\int_0^t \frac{f(t-p)}{p^{3/2}}e^{-x^2RC/4p}\,dp.$$

For given values of R, C, and x, this integral is easily evaluated *numerically* (see the *MATLAB* code **accon.m**), as a function of t, even for an $f(t)$ so complicated that being asked to do it analytically would have struck horror in the throats of Fourier and Thomson.

To back-up the claim made in the final sentence of the previous shaded box, the code **accon.m** (for *A*tlantic *C*able *CON*volution) shows how to find $v(x, t)$, with *MATLAB*, for just about any $f(t)$. (The values of x, R, and C are those used in Fig. 5.2.) The logic behind the code is as follows:

1. start and stop values of time are specified, along with the number of points desired in the final plot of $v(x, t)$;
2. from (1), *MATLAB* creates a vector of time values;
3. for those time values, the input function $f(t)$ is evaluated;
4. for those time values the function $h(t) = \frac{e^{-x^2RC/4t}}{t^{3/2}}$ is evaluated;
5. *MATLAB*'s built-in *conv* function performs the convolution of $f(t)$ with $h(t)$, and scales the result by the factor $\frac{1}{2}x\sqrt{\frac{RC}{\pi}}$.

```
%accon.m/created by PJNahin for Electrical Transients (11/9/
2017)
tstart=1e-6;tstop=10;n=10000;x=2000;
R=3;C=5e-7;deltat=(tstop-tstart)/n;
t=tstart:deltat:tstop;
g1=x*x*R*C/4;g2=0.5*x*sqrt(R*C/pi);
for k=1:length(t)
  if t(k)<5
    input(k)=1;
  else
    input(k)=0;
  end
end
input=input*deltat;
h=exp(-g1./t)./(t.^1.5);
v=conv(input,h);
v=g2*v;
plot(t,v(1:length(t)),'-k')
xlabel('time (seconds)')
ylabel('volts')
```

To test the code, an input of all 1's (to simulate a unit voltage step input) is used. When run, the code's result is, to the eye, an exact replica of a plot of the theoretical result given in (5.17) and (A3.21).

Now, to do a problem that Fourier and Thomson would *never* have even attempted analytically, suppose

$$f(t) = \begin{array}{cc} \sin\left(t\ sin\left(t^2\right)\right), & 0 < t < 5 \\ 0, & t > 5 \end{array}$$

All that is required in **accon.m** to handle this $f(t)$ is the replacement of the line.

```
input(k)=1;
```

with the lines[10]

[10]This sort of function, in which the instantaneous frequency changes with time, is an example of FM (frequency modulation), used in such important electronic gadgets as radio and pulse compression radar. FM, and its gadgets, would have been magic to Fourier and Thomson.

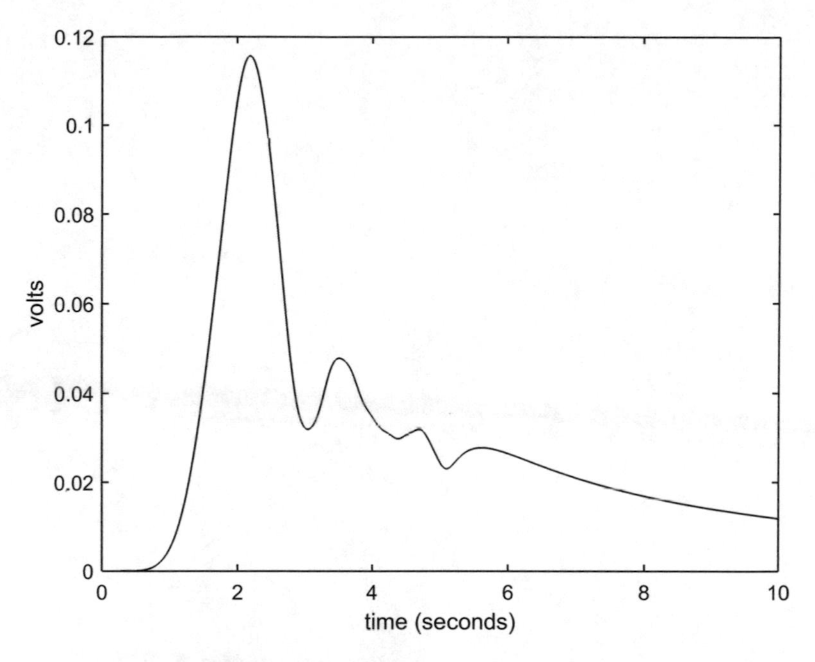

Fig. A3.1 The transient response of the Atlantic Cable to the input $f(t) = \sin(t\, sin\,(t^2))u(5-t)$

```
if t(k)<5
  input(k)=sin(t(k)*sin(t(k)^2));
    else
  input(k)=0;
  end
```

The result is shown in Fig. A3.1.

Appendix 4: A Short Table of Laplace Transforms and Theorems

Transform pair	First discussed at
$u(t) \leftrightarrow 1/s$	(3.11)
$e^{-at}u(t) \leftrightarrow \frac{1}{s+a}$	(3.12)
$\cos(\omega_0 t)u(t) \leftrightarrow \dfrac{s}{s^2+\omega_0^2}$	(3.15)
$\sin(\omega_0 t)u(t) \leftrightarrow \dfrac{\omega_0}{s^2+\omega_0^2}$	(3.16)
$t^n u(t) \leftrightarrow \frac{\Gamma(n+1)}{s^{n+1}}$	(3.36)
$\frac{1}{\sqrt{t}}u(t) \leftrightarrow \sqrt{\frac{\pi}{s}}$	(3.44)
$\sqrt{t}u(t) \leftrightarrow \frac{1}{2s}\sqrt{\frac{\pi}{s}}$	(3.45)
$J_0(at)u(t) \leftrightarrow \frac{1}{\sqrt{s^2+a^2}}$	(3.47)
$I_0(at)u(t) \leftrightarrow \frac{1}{\sqrt{s^2-a^2}}$	(3.48)
$\delta(t) \leftrightarrow 1$	(3.52)
$\left\{1 - erf\left(\frac{a}{2\sqrt{t}}\right)\right\}u(t) \leftrightarrow \frac{1}{s}e^{-a\sqrt{s}}$	(3.79)
$\frac{ae^{-\frac{a^2}{4t}}}{2\sqrt{\pi t^3}u(t)\leftrightarrow e^{-a\sqrt{s}}}$	(3.85)
$I_0\left(p\sqrt{t^2-k^2}\right)u(t-k) \leftrightarrow \frac{e^{-k\sqrt{s^2-p^2}}}{\sqrt{s^2-p^2}}$	(5.33)
$f(t)u(t) \leftrightarrow F(s)$	(3.1)
$tf(t)u(t) \leftrightarrow -\frac{dF(s)}{ds}$	(3.8)
$\dfrac{f(t)}{t}u(t) \leftrightarrow \displaystyle\int_s^\infty F(x)dx$	(3.9)
$\frac{df(t)}{dt}u(t) \leftrightarrow sF(s) - f(0+)$	(3.2)
$\frac{d^2f(t)}{dt^2}u(t) \leftrightarrow s^2F(s) - sf(0+) - f'(0+)$	(3.5)
$\frac{d^3f(t)}{dt^3}u(t) \leftrightarrow s^3F(s) - s^2f(0+) - sf'(0+) - f''(0+)$	(3.6)
$\displaystyle\int_0^t f(x)dx \leftrightarrow \dfrac{F(s)}{s}$	(3.7)

(continued)

P. J. Nahin, *Transients for Electrical Engineers*,
https://doi.org/10.1007/978-3-319-77598-2

Transform pair	First discussed at
$e^{-at}f(t) \leftrightarrow F(s+a)$	(3.13)
$f(t-t_0)u(t-t_0) \leftrightarrow e^{-st_0}F(s)$	(3.14)
$\lim_{s\to 0} sF(s) = \lim_{t\to\infty} f(t)$	(3.18)
$\lim_{s\to\infty} sF(s) = \lim_{t\to 0+} f(t)$	(3.19)
$f(t) * g(t) \leftarrow F(s)G(s)$	(3.87)

Index

A
Abel, N.H., 72
Ampere, A.M., 1
Arcs, xii, 20, 21, 31, 37
Atlantic Cable, 93, 133, 136, 137, 140–143, 160, 169–181

B
Bernoulli, D., 169, 170
Bernoulli, J, 171
Bessel function, 73, 74, 149
Bessel, F.W., 73
Binomial theorem, 72

C
Capacitors, xii, xiii, xv, xvi, xviii, 1, 2, 4, 5, 7, 8, 11, 18, 19, 21, 25, 26, 29, 31, 32, 37, 43, 61–63, 77, 79, 89, 101, 107, 111, 120–122, 130
Churchill, R.V., xvi, 99
Clamping, 123–125
Conductance, 134, 136, 166
Conjugate (complex), 12
Conservation (of electric charge), xiii, 4, 26
Conservation (of energy), 4, 18, 90
Constant current source, 1
Convolution, 93–100, 102, 103, 160, 179
Coulomb, C., 4
Cramer, G., 16

D
Delta function, xviii, 75
Dirac, P., xviii, 76
Doetsch, G., 100
Dummy variable (of integration), 5, 53, 57, 69, 173, 176

E
Equation (algebraic), 12, 51, 108
Equation (differential), xi, xv, xix, 5, 6, 51, 73, 93, 100, 133, 137, 140, 168, 169, 171
Equation (separable), 32
Equation (telegraphy), 135–137
Error function, 69, 71, 93–100, 143, 177
Euler, L., 68, 161
Even function, 175
Exponential functions, 56, 73

F
Factorial function, 68
Faraday, M., 2
Final value theorems, 60, 81, 100, 152, 159
Flux (magnetic), xiv, 2, 8, 9, 11, 20, 21, 26, 90
Fourier, J., 93, 137
Frequency (natural), 22–24, 41, 116
Functions, xviii, xix, 11, 16, 18, 25, 51–53, 55–61, 63, 67, 68, 71, 73–76, 80, 85, 93, 97, 98, 100–102, 104, 112, 133, 149, 150, 153, 156, 157, 160, 163, 169, 170, 172, 174, 175, 179, 180

P. J. Nahin, *Transients for Electrical Engineers*,
https://doi.org/10.1007/978-3-319-77598-2

G
Galvanometer, 143
Gamma function, 68

H
Heaviside, O., xv, 55, 76, 101, 153
Henry, J., 2
Hertz, H., 22
Hurwitz, A., 129
Hyperbolic functions, 155

I
Ilaplace (MATLAB), 87
Impedance, 101, 125, 130, 158, 159, 166, 168
Impulse functions, xvii–xix, 18, 75, 102
Inductance (mutual), 21, 117, 134
Inductance (self), 3, 21, 89
Inductors, xii–xvi, xviii, 1, 2, 4, 7, 8, 11, 14, 20, 21, 25, 31, 37, 43, 45, 90, 101, 105, 107, 111, 116, 117, 125, 130, 133
Initial conditions, xvii, 45, 154, 171
Integration-by-parts, 102

J
Jacobians, 70
Joule, J., 7

K
Kelvin, L., 16, 137
Kirchhoff, G.R., 4

L
Laplace transforms, xvi–xx, 42, 51–105, 120, 125, 128, 137, 138, 140, 141, 147, 150, 152, 169–183
Law of squares, 144
Leibniz, G.W., 5
Lenz, H., 90
Loading (inductive), 146, 153
Loop currents, xv, 78

M
Magnetic coupling, 21, 41–49, 116
Maxwell, J.C., 78
McShane, E., 77
Morse code, 145

N
NE-2 gas bulb, 120
Newton, I., xviii, 137

O
Odd function, 175
Ohm, G., 2
Operator (resistance), 101, 111, 166
Oscillator (relaxation), 122

P
Paradoxes, 7
Partial fraction expansions, 65, 82, 85, 87, 106, 107, 123, 125, 127, 129
Phasors, 130
Power (electrical), 11
Power (function), 98

R
Ramp function, 67
Rayleigh, L., 142
Resistors, xii, xiv, xv, xviii, 1, 2, 7, 8, 10, 11, 13, 29, 61, 79, 83, 88, 89, 91, 101, 102, 107, 111, 112, 120, 123, 126, 127, 129, 130, 132, 133
Routh, E., 129
Routh-Hurwitz algorithm, 129

S
Sampling property (of the impulse function), 76
Schwartz, L., xviii, 77
Shifting theorem, 57, 59
Siemens, E.W., 134
Sinusoid function, 55–61
Sobolev, S., 77
Square-wave, 58–60
Steady-state, 13, 16, 37, 61
Step function, 55, 58
Strutt, J.W., 142

T
Thomson, W., 16, 17, 169, 172
Tolstov, G., 173
Transformer, 1, 87, 105, 112
Transform (Fourier), 169, 172, 173, 177
Transform (Laplace), 5, 11, 21, 31, 42, 51–104
Transmission line (finite), 133

Transmission line (infinite), 176
Transmission line (leakage-free,
non-inductive), 178
Transmission line (lossless), 157, 159

V
Varignon, P., 171
Victoria (Queen of England), 137
Volta, A., 1

W
Watt, J., 7
Wave (incident), 157
Wave (reflected), 157

Z
Zener diode, 122

Printed in the United States
By Bookmasters